333.7 J

£8·95

The Greenhouse Effect

and Ozone Layer

Philip Neal

Dryad Press London

Contents

Acknowledgments

The author wishes to thank the following for their help in the preparation of this book. David Vodden, Rita Neal, John Letcher of Farm Gas Ltd, Bishops Castle, Australian Glaciation Survey, the Meteorological Office Library, Environment Canada and UNEP.

The Author and the Publisher thank the following for their permission to reproduce photographs. Australian High Commission, page 50; Museum of London, page 52; BBC Hulton Picture Library, pages 6, 7; Popperfoto, Paul Popper Ltd page 43 (bottom); Norwegian Embassy, page 57; C. J. Gilbert, British Antarctic Survey, page 38 (bottom left); 45; International Centre for Conservation Education, page 28; Tilda Rice Ltd, page 13, 49 (top right); CEGB, page 26; Sheffield City Library, page 30; Meteorological Research Flight, page 38 (top lef); Johnson Matthey Chemical Ltd, page 43 (top); Meteorological Office Library, page 25, 38 (top right), 39 (bottom left); CEGB NW region, page 35; Friends of the Earth, 47; Farm Gas Ltd, page 49 (bottom); Intervention Board for Agricultural Produce, page 8, 23; United Nations Environment Programme (UNEP), page 31; British Brown-Bouverie, page 33; William J. Neal, page 11 (bottom); United States Air Force, page 59; Carol Chapman, page 49 (top left); Batsford Library cover (top) and page 9.

The diagrams on the cover and pages 4, 5, 18 (top), 19, 23, 25, 29, 35, 48, 51, 53 (bottom), 55 and 59 are by Sue Prince; page 45 from UNEP; page 31 graph and 45 map from Environment Canada; page 32 Electricity Council.

All other maps, diagrams and photographs are those of the author.

Cover illustrations
Top: Drought affected land (Batsford library)
Centre: Factory chimney, West Midlands (Philip Neal) *Below left:* BBC weather forecast (Philip Neal) *Below Bottom right:* CFCs propel aerosol spray (Sue Prince)

Title page: Power station (Philip Neal)

© Philip Neal 1989
First published 1989

Typeset by Latimer Trend & Company Ltd, Plymouth
and printed by
The Bath Press, Bath
for The Publishers
Dryad Press,
4 Fitzhardinge Street,
London W1H 0AH

ISBN 0 85219 822 1

Introduction

Ask people who have lived in the UK for a long time, perhaps your grandparents, if they think that the summer now is less sunny than it was in the times when they were children. I am sure I know what the answer will be. Even allowing for the fact that we all look back on the years which have passed with happiness (we say that we look "through rose-coloured glasses"), it is true to say that fifty years ago the summers were sunnier and, in particular, "flaming June", July and August were times of hot days, with nights warm enough to make sleep uncomfortable. Are we just going through a temporary change of climate, or is it permanent?

Certainly, in other parts of the world, strange things are happening to the weather. In summer 1988, for 54 days and 54 nights, more than half of the people of the USA endured the worst heatwave since scientists started officially to record temperatures. New York's climate was similar to that of the tropical Amazon region. Newspapers stopped bothering to report temperatures that topped the 90°F mark. The *New York Times* wrote: "The suns and seas and sins of man have combined to transfer New York life into a seemingly endless slog through simmering broth."

The broth was hot, pollution-laden air, thick with vehicle exhaust fumes and industrial haze. The excessive heat hastened the rotting of the usual piles of rubbish on the beaches nearby, and unfortunately, this rubbish included infected hospital waste discharged into the Hudson River. The stench was so bad that no cooling sea-bathing or relaxing on the shore was possible.

Why are such changes occurring in the climate of places around the world? Is there a cause which we can isolate? The popular explanation is called THE GREENHOUSE EFFECT. This book explains what this is, what part of it is accounted for by natural causes and what has been caused by people polluting the air. Why is this air pollution occurring and what can be done to stop it?

Other happenings are taking place in the atmosphere which surrounds our planet. One, which concerns "holes" in the OZONE LAYER, is of particular interest. Examples of unusual weather around the world are described and some other explanations investigated.

The main purpose of the book is to help you decide if people and their modern ways have pushed nature too far. How should we dispose of our waste? Ought we to burn fossil fuel to make electricity? Is it right to burn down the rainforests? Are governments spending enough money on research into air pollution? These and a host of other questions need to be investigated, for the "sins of man", as the *New York Times* put it, are important matters for all of us considering conservation.

The atmospheric pollution broth

The various human activities of a typical industrial urban area give rise to many chemicals which join together in the air above to form a complex brew of pollution. Some of these chem-

As (arsenic) from coal, oil burning, glass making: causes lung, skin cancer

Cd (cadmium) from coal, oil, waste burning, metal smelting. Damages lungs, kidney, weakens bones

CFCs (chlorofluorocarbons) from aerosols, industry, refrigeration: absorbs heat, destroys ozone

SO_2 (sulphur dioxide) from coal, oil burning, metal smelters, fossil fuel power stations: main cause of acid rain, hurts eyes, lungs

SiF_4 (silicon tetrafluoride) from chemical factories: hurts lungs

Pb (lead) from vehicles, metal smelters: stunts growth, damages brain, causes high blood pressure. Added to petrol—"unleaded" now on sale

PAN (peroxyacetyl nitrate) from NO_x and HC photochemically: hurts eyes, lungs

OH (hydroxyl radical) from NO_x & HC in sunlight: adds to acid rain

O_3 (ozone) from NO_x and HC photochemically: absorbs heat: hurts eyes, lungs: Protects people from ultra violet rays by forming shield in upper air

NO_x (nitrogen oxides) from motor exhaust, coal, oil burning; some formed photochemically: makes ozone, causes breathing problems

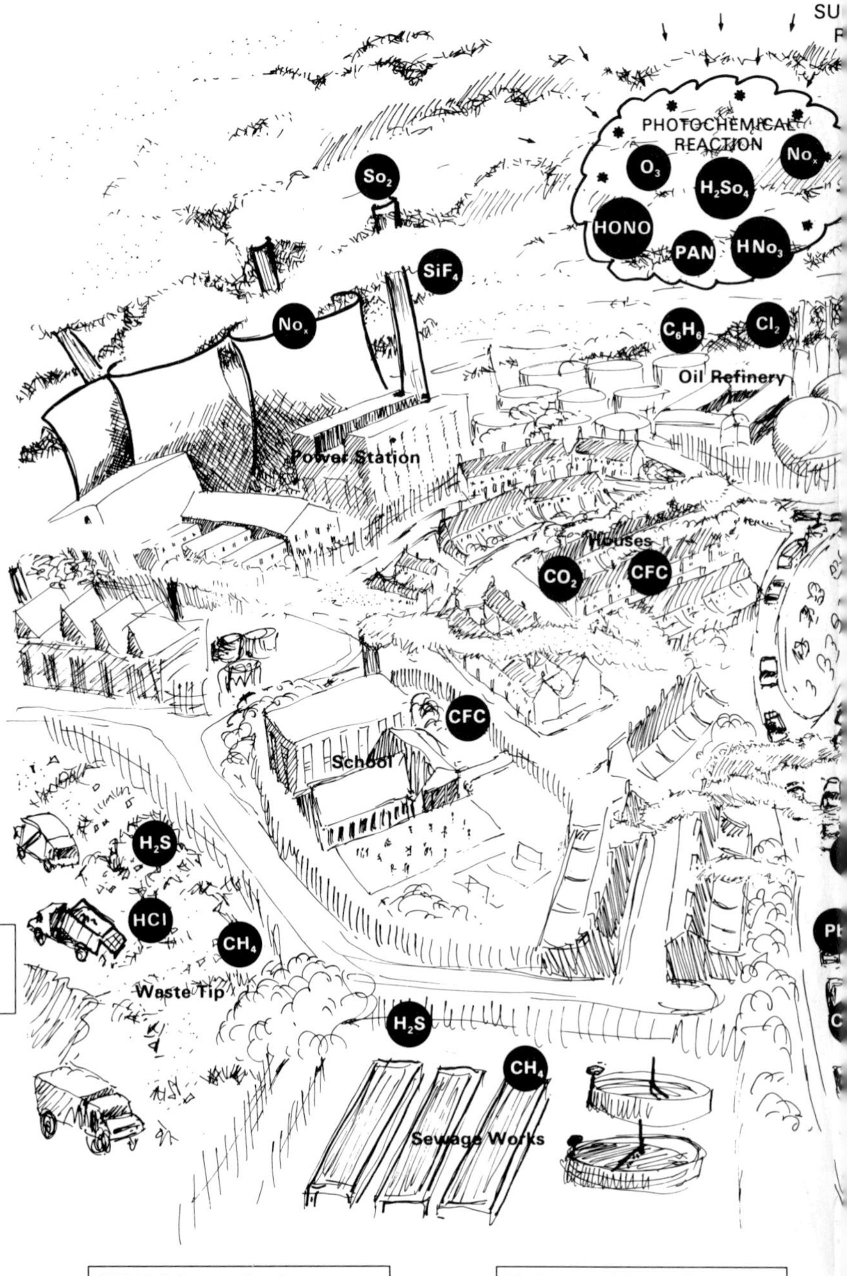

Ni (nickel) from metal smelters, fossil fuel burning: can cause lung cancer

Mn (manganese) from steel making, power stations: one cause of the body-shaking Parkinson's disease

PHOTOCHEMICAL REACTION

O_3 NO_x H_2SO_4 HONO PAN HNO_3

SO_2 SiF_4 NO_x C_6H_6 Cl_2

Oil Refinery

Power Station

CO_2 CFC Houses

CFC School

H_2S HCl CH_4 Waste Tip

H_2S CH_4 Sewage Works

icals come directly from definite sources: for instance SO_2 from the chimneys of coal-burning power stations. Others are made indirectly by a photo-chemical reaction triggered by the sun's rays. Ozone is formed by the action of sunlight on nitrogen oxides, amongst other chemicals.

This page shows just a few of the sources of the harmful chemicals which concentrate together to make the simmering broth referred to in the introduction and briefly tells you something of the dangers they cause.

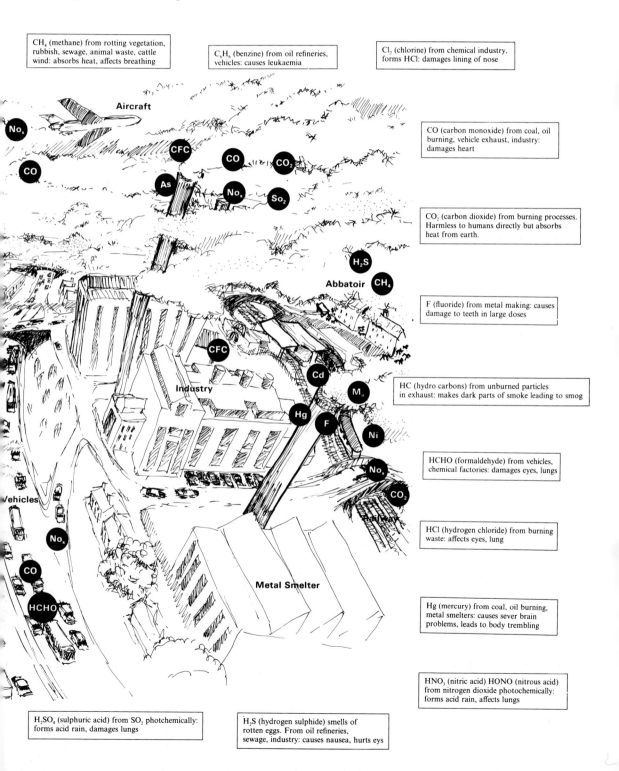

CH_4 (methane) from rotting vegetation, rubbish, sewage, animal waste, cattle wind: absorbs heat, affects breathing

C_6H_6 (benzine) from oil refineries, vehicles: causes leukaemia

Cl_2 (chlorine) from chemical industry, forms HCl: damages lining of nose

CO (carbon monoxide) from coal, oil burning, vehicle exhaust, industry: damages heart

CO_2 (carbon dioxide) from burning processes. Harmless to humans directly but absorbs heat from earth.

F (fluoride) from metal making: causes damage to teeth in large doses

HC (hydro carbons) from unburned particles in exhaust: makes dark parts of smoke leading to smog

HCHO (formaldehyde) from vehicles, chemical factories: damages eyes, lungs

HCl (hydrogen chloride) from burning waste: affects eyes, lung

Hg (mercury) from coal, oil burning, metal smelters: causes sever brain problems, leads to body trembling

HNO_3 (nitric acid) HONO (nitrous acid) from nitrogen dioxide photochemically: forms acid rain, affects lungs

H_2SO_4 (sulphuric acid) from SO_2 photchemically: forms acid rain, damages lungs

H_2S (hydrogen sulphide) smells of rotten eggs. From oil refineries, sewage, industry: causes nausea, hurts eys

Aircraft

Abbatoir

Industry

Vehicles

Metal Smelter

Railway

The "Dust Bowl" again

The soil of the US state of Oklahoma blows away as a dust-storm. This was the ultimate effect of the drought of the 1930s and gave rise to the name of DUST BOWL for this part of the USA.

The American author John Steinbeck wrote *The Grapes of Wrath*, and from this book an equally well-known film was made. The story tells of farmers and their families in Oklahoma, a mid-western state of the USA, during the drought of the 1930s. Such was the dryness of fields, bare of any crops because of the lack of rain, that the soil blew away as dust and the "Dust Bowl" region was created. People left the area and travelled as best they could towards the west and, in particular, to California, seeking new ways of earning a living. The way in which the then President of the USA, Franklin D. Roosevelt, tackled the problem, creating the Tennessee Valley Authority with its dams, hydro-electricity and irrigation schemes, is another story, but one which is well worth investigating from your library books.

Fifty years later, President Ronald Reagan had a similar problem to tackle. In July 1988 he visited Illinois, another of the central states of the USA. He told farmers: "This is the worst natural disaster since the dust bowl of the 1930s." He had gone to look at the results of a devastating drought. Soaring temperatures accompanied by an almost complete lack of rain across the whole of the west of the country, led to fields, which ought to have been full of ripening corn (we call it maize – you may know it as "corn on the cob" – which makes the crisp cornflakes of breakfast time), having half-grown and shrivelled stalks in them. President Reagan, who was raised in Illinois, told reporters: "I know how high they should be because I used to hide in them as a kid." Stalks which should have been two metres tall were barely half that height.

The grain-growing areas of North America are seven times bigger than the whole of the British Isles. Oklahoma is in the south-west of the region.

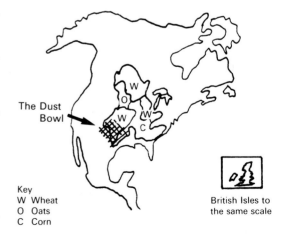

The Dust Bowl

Key
W Wheat
O Oats
C Corn

British Isles to the same scale

Throughout the summer, from the Rocky Mountains in the west to Lake Michigan in the east, from the state of Alabama in the south to the Canadian border and beyond in the north, the sun beat down pitilessly on the farming areas of the mid-west. Temperatures were relentlessly above the 100°F level for days on end. Even across the Rocky Mountains, where the nearness to the Pacific Ocean normally keeps the high temperatures bearable, the mercury in the thermometers soared above 100° on many days. On 17 July San Francisco had the hottest day since records of the weather had been kept. In Sacramento it reached 120°F and so much demand was put on the electricity supply for air-conditioning systems in houses, shops and offices, that power cuts occurred.

News stories emerging from this extraordinary climatic situation emphasized the problems of carrying on with normal living. Chicago's beaches, along the shores of Lake Michigan, were left empty at the peak of the holiday season, because the sand became too hot to walk on and in-shore waters too uncomfortably warm for swimming. Iced water machines had to be placed in the streets to provide cool liquid for passers-by and so prevent body dehydration. (Not having enough water in your body causes ill effects rather like an attack of influenza.) At Philadelphia Zoo they had to buy nearly half a tonne of ice a day, in order to cool down the polar bears, gorillas and tigers.

Performing the usual activities of daily life became unbearable.

June 1938. Fifty years ago this family packed up all they could and set out to walk away from the desolation of the Oklahoma countryside to try to find a better home somewhere to the west.

Drought – North America, 1988

As a result of the drought, American and Canadian grain-growing farmers lost at least a third of their corn (maize) and wheat crops. For the first time in fifteen years, the grain harvest was less than 200 million tonnes in the USA and down by nearly half in Canada, to about 17 million tonnes. World demand for grain is enormous. The Worldwatch Institute, based in Washington, estimated that, in 1988, 152 million tonnes more grain was consumed than had been produced. At the moment, there are sufficient stockpiles of grain to meet the need. But how long would the stockpiles last if a continuing alteration in the climate led to continuing decreases in the amount of grain produced?

What was true for grain was also true for the crops of soya beans, tobacco and cotton. Thunderstorms late in July brought a little hope that the later crops might produce some harvest, but for many farmers of the mid-west, the amount of money which they might earn was less than the cost of the seed which they put into the ground in the first place.

Farm animals were also affected by the heat and drought. Farmers had enormous problems in supplying drinking water to their cattle and pigs to prevent their dying of heatstroke. Two months of rainless weather meant that grass did not grow as it should have done. The cattle ate what was available, and little was left to be made into hay for winter feed. Farmers were forced either to slaughter half their herds or to buy in feed for their animals at higher prices than normal because of the shortage and the high demand. So many animals going to the butchers caused the price of meat to fall, and this brought some ranch owners to the brink of financial ruin. Help from the US government came through the

In normal times the grain belt of the USA and Canada produces so much that the surplus has to be put into grain stores and silos.

Drought Assistance Act, signed by President Reagan, but money to help could not build up the stocks of grain or the herds of cattle, so that the future of farming in the mid-west will be affected for many years to come. Yet, in spite of this, many grain-growers came out of 1988 with a large profit. Money from farm insurance, sales of 1987 grain stocks at high prices, and the government subsidy brought in more cash than if the drought had not taken place. Such are the economics of modern agriculture.

Newspaper reports told of people crowding into shopping centres on a scale normally reserved for the Christmas rush, as they found the air conditioning more attractive than the outside air. While this may be slightly amusing to read, it is not so funny to consider what might be happening. Will it be better in the future? Is this drought the start of a new pattern of weather?

Soil without enough moisture becomes parched. It shrinks into blocks with large cracks between them, and is completely useless for crop growing.

Cars off the road

During the drought, over twenty large cities in the affected area asked drivers not to use their cars. Since there was no petrol shortage, there must have been some other good reason for the directive. The explanation was an attempt to cut down the polluting exhaust gases coming from the engines.

DUST BOWL BLAMED ON GREENHOUSE EFFECT.

This was the heading above one story about the farming crisis in America. A scientist from NASA (the National Aeronautics and Space Administration) triggered the report when he told a committee of the US Congress that the "dust bowl drought" was 99 per cent certain to be caused by pollution of the air and that it was an early sign of a dangerous change in climate, a change which leads to a general warming of the atmosphere around the world. This, in turn, affects the rainfall, as well as having other side-effects. He said this atmospheric pollution was called THE GREENHOUSE EFFECT. Was this the real cause of the drought?

**THE GREENHOUSE EFFECT causes a warming
of the atmosphere around the earth.**

The inland shore – the Sahel

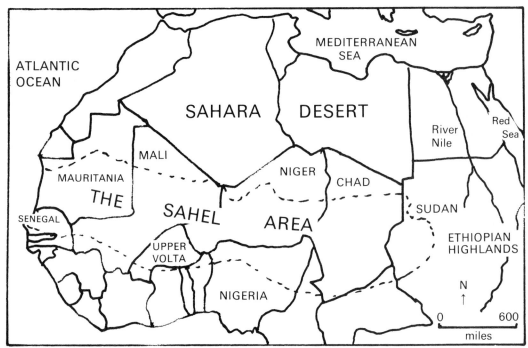

North Africa: the Sahel area.

Seas have shores, areas of land beside the sea over which the water ebbs and flows as the tide falls and rises. Deserts, too, have border lands which are sometimes covered in vegetation and sometimes completely barren, depending on the changes in weather from year to year. In Africa, south of the Sahara Desert, they call the 5000-kilometre-long borderland "The Shore": in the Arabic language, "The Sahel".

A thousand kilometres wide in places, this semi-desert territory has been home to wandering nomadic peoples, the most famous of whom were the Tuaregs. I say *"were"*, because that nomadic way of life has all but disappeared, along with much of the Sahel. As if the sea level had risen and covered the shore, the sands of the Sahara have moved south to cover much of this land. The term for this happening is "desertification", about which you can read much more in the book *The Encroaching Desert* in this series.

The reasons for the change in the Sahel are many and complicated. If you look at the map, you will see that the Sahel stretches from the Atlantic coast in Senegal, through Mali, to the Sudan. The higher lands of Ethiopia lie beyond. Some of these place names must ring a bell in

your minds; perhaps if I say "Bob Geldof" and "Band Aid" it will bring home to you that these are the areas involved with the horrifying starvation events of the past few years.

Many of the reasons for the catastrophe are political, from destructive civil wars to well-intentioned development programmes which went wrong. Many reasons are social, to do with a breakdown of the old tribal systems, where people lived within the limits set by nature rather than trying to override the natural conditions of soil, water supply and available vegetation. Some reasons are to do with the climate, and these are the ones which interest us here.

Droughts have always plagued the area; otherwise it would not have been a desert shore, a place of periodic change. However, the droughts of recent times have been more frequent and longer. The worst lasted from 1969 to 1974, five long, dry and hunger-making years. After some recovery to "normal" conditions, further drought in 1984 and 1985 had a crippling effect, as we all well know, and tried, through our purses, to "aid".

Droughts are the extreme of dry periods. Even where conditions are near to the normal of

10

Thirsty and starving cattle wander in the drought-stricken Sahel.

the past hundred years there is evidence that the climate has been changing slowly but steadily. Between 1920 and 1960 the average annual rainfall in parts of Mali was 270 mm ($10\frac{1}{2}$ inches). During 1961 to 1984 this meagre amount dropped to 200 mm (under 8 inches). In the UK, most of the rain which falls remains available for our needs. It is estimated that in the Sahel, by contrast, almost 90 per cent of the rain evaporates because of the burning sun. In addition to there being below-average rainfall coming into the area, it is believed that a spreading desert in itself encourages less rainfall. Places covered with vegetation act as a sort of reservoir, and the water which is released into the air by the plants condenses to form rain-making clouds, a natural cycle broken by the barren land.

For whatever reasons, the Saharan shore has shifted 100 km (60 miles) to the south in the last forty years. At the extreme eastern end of the Sahel is Ethiopia. Here, too, famine has been a major problem, as drought conditions have prevailed. Yet, in 1988, the problem was one of flood. Excessive rains in the Ethiopian Highlands, whose streams and rivers feed the waters of the Blue and White Niles, led to vast tracts of land being submerged, especially in the area around Khartoum, where the two rivers join to form the Nile itself. Higher areas always receive more rain than lower places, but the amounts of 1988 were greatly above the norm and were another example of the climatic disturbances worldwide.

One explanation put forward for the difficulties of the Sahel and Ethiopia is The Greenhouse Effect. Is this the cause of desert and flood?

As it used to be

The Tuaregs were adapted to living with nature. Under the laws of Islam, they were allowed to use certain ways through the Sahel only a certain number of times each year. The practical effect of this was to prevent overgrazing, which is one of the most severe ways of causing the spread of the desert. With people encouraged by their government to settle in villages their animals repeatedly graze from the same land. When the rains failed, the plant life was completely eaten away, instead of being left to recover in the course of time.

As it used to be. *Nomadic camel breeders moved camp regularly and so conserved the limited vegetation.*

Monsoon failure

One week the land is parched, cracked and bone-dry. The fields are brown, dust-storms rasp the throat and fog the view, and electric power fails as the water behind the hydro dams is too low to activate the turbines. Lethargic cattle wander, their ribs outlined against taut skins. Sellers overcharge for buckets of drinking water and refuse the pleading of those without a rupee to their name.

The next week seas of mud impede the traffic. Fields rapidly turn to growing green, torrential rain beats down incessantly, reservoirs fill quickly to overflowing, drowned bodies of cattle float by on the swollen rivers, and the water sellers now peddle umbrellas.

The monsoon has broken: the wet season has arrived. Such is the story of much of the land of southern and south-eastern Asia before and after the wet monsoon. In India the event is due in the first weeks of July; sometimes it is early, sometimes it is late, but inevitably it comes, bringing welcome relief from the drought but leading on to constant discomfort and wet. "Inevitably" was not the right word, however, in 1987, when the monsoon failed in some parts. Nor did the rain come "inevitably" in 1982, when the monsoon failed in certain other places.

Five years between two such unusual incidents is too close for comfort, even though different places were involved. What can be the cause of the monsoon's repeated failing in this way?

Most regions experience two monsoons each year, one dry, the other wet. Monsoons occur because the enormous land mass of Asia is surrounded by an equally enormous water mass of the Pacific and Indian Oceans. In January, the heartland of Asia is bitterly cold, while the seas around, much of them within the tropical zone, are very warm. Above the warm seas the air rises, creating a low-pressure area at the surface. Above the land the cold air falls, to make an area of high pressure. Low pressure occurs over the water, high pressure over the land. Air moves from high to low, out from the continent across the oceans. The moving air – the wind – is dry. As it moves across the fields, it draws away what moisture there is.

In summer the situation is reversed. The heartland of Asia heats rapidly to high temperatures. The wind travels from the oceans, across the land, bringing water-laden air which condenses out as rain. The wet monsoon breaks.

The basic conditions do not change from year to year. So why do the monsoon rains some-

The word "monsoon" is thought by some people to mean "rain" and by others to mean "wind". In fact, the Arabic derivation of the word is MAUSIM, meaning "season". In some places, particularly the islands, both the summer and winter winds cross the sea at some point, resulting in a double wet monsoon every year.

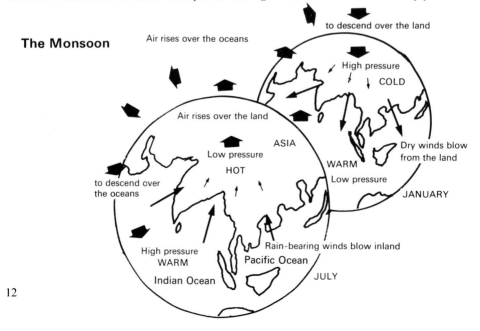

The Monsoon

Air rises over the oceans

to descend over the land

High pressure

COLD

Air rises over the land

ASIA

Dry winds blow from the land

Low pressure

HOT

WARM

Low pressure

to descend over the oceans

JANUARY

High pressure
WARM

Rain-bearing winds blow inland

Pacific Ocean

Indian Ocean

JULY

The farming of Asian lands is based on the annual cycle of monsoon conditions. A failed monsoon can lead to farming failure. The need for much rainfall is obvious from this picture of rice paddy terraces in SE Asia.

times fail? One theory recognizes that just a small rise of temperature over the Western Pacific Ocean can have a very large effect on climatic conditions. The wet monsoon results from moisture-laden air travelling from the ocean over the land, where it is forced to rise and therefore to cool. The cooler air cannot retain its moisture, which condenses as cloud and later as torrential rain: the monsoon arrives. However, if clouds form over the ocean in summer, they hold the water and prevent it travelling over the land. The critical temperature for clouds to form over the Western Pacific Ocean appears to be 28°C. As long as the thermometer stays below this level, clouds are not formed over the sea. But if, instead, a small rise above 28°C takes place, the warmer air rises and cools and massive clouds are formed. The water "intended" for the land is trapped. These were the conditions observed in 1987.

The critical point to notice is that it takes only a small rise in temperature to upset the monsoon pattern. And a small rise in temperature is being blamed on the Greenhouse Effect. Is this the cause of monsoon failure?

Despite the monsoon rain in India, planting rice goes on in the paddy field, by women ankle-deep in mud.

British weather!

On October 15th, 1987, a hurricane hit southern Britain. Fifteen million trees were uprooted. Roofs were blown off buildings. Luckily it occurred at 4.00 a.m. when people were in bed. Otherwise thousands would have been killed or injured. At Sevenoaks in Kent, six of the seven oak trees were felled by the wind.

The weather experts at the Meteorological Office said that it was not an actual hurricane. If so, I would not wish to experience a real one. The screaming wind and frightening storm of 15th October, 1987, were bad enough. Trees crashing, roof tiles flying, windows shattering and walls falling were incidents ferocious enough for me to call the storm a hurricane.* Never before in recent history had such a severe event occurred in England. Even a year later the south of England was littered with fallen trees, and many roofs were still covered with tarpaulins.

The year that followed the storm was odd, to say the least, as far as the weather was concerned. Where was the summer of 1988? A brief spell of fine weather in June, especially in Scotland, was followed by days of rain and the need for extra heating in homes and offices. July and August recorded all-time highs of rainfall. The major sporting events of Wimbledon tennis and Lords Test Match cricket were affected. The Open Golf championship was played in a deluge. And all of this following a warm winter when the snows did not come and, throughout Europe, winter skiing was impossible on slopes which remained green.

Wet weather in the UK is usually brought about by low-pressure areas moving from west to east across the country, from the Atlantic Ocean to mainland Europe. Within a low-pressure system, cold and warm air meet at a junction called a "front" and rain is formed. In normal summers high pressure coming from the warmer Tropics to the south blocks these low-pressure areas. They are forced to move north around the High, thus avoiding Great Britain. But not in 1988. Something seemed to have affected the major pattern.

The popular explanation for the change in the weather pattern was that it was caused by the worldwide Greenhouse Effect. This would result in a warmer winter, but more rainfall, extra cloudiness blocking out the sunshine, to make the summer cooler than normal over most of Britain.

What, in fact, do scientists speculate about Britain and the British climate if the temperature does go up by even a few degrees? They say that the southern part of England may take on

*It is true that the winds were generally below 100 mph. In September 1988 the winds of Hurricane Gilbert in the Gulf of Mexico reached a record 200 mph.

IN?

OUT?

Kent is the nearest English county to France. It is full of apple and plum orchards, hop gardens (hops are used to make beer) and strawberry fields. With global warming, will orange groves replace the orchards, and wine-making from vineyards replace beer-making and hop gardens?

the weather conditions of the southern parts of France, with warm and wet winters and hot, dry summers. Vine-growing will be possible, with growers able to produce red wine grapes to rival those of the Bordeaux region in France. It is probable that a higher average temperature will mean more extremes of weather, with storms, similar to the "hurricane", flooding, and harsher frosts in the winter. The side-effects will be interesting. Conifer trees will be more difficult to grow, but the "new" climate will help broad-leaved trees. Our fruit crops will be peaches and nectarines rather than apples and plums. It will not be possible to grow daffodils and crocuses easily, except in the north. We will gain butter-flies and lose birds, particularly those which live in the higher lands and in the north of Scotland. Skiing will become impossible in Scotland's Aviemore. As for the coast, a rising sea level will make the rebuilding of coastal defences essen-tial, not only to prevent flooding but to prevent the salting of the soil from the sea water.

Will all of this make living in Britain better or worse? Impossible to answer, until it happens – if it happens. What is the real cause of the apparent change in the British weather? Is it the Greenhouse effect?

Newspaper weather forecasts are either seriously presented statistics and forecasts or have a more "popular" cartoon approach. Whichever method is chosen, every paper has its weather section daily.

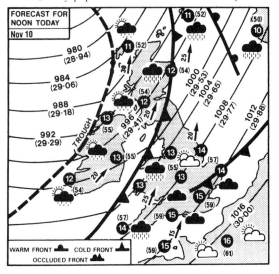

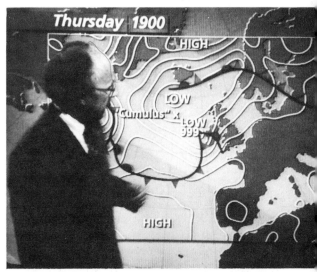

Weather in Britain is always one of the main topics of conversation. "What was the weather like?" is the first question we ask someone on their return from holiday. Newspapers, the BBC and ITV give us regular forecasts based on the observations of the Meteorological Office. They make much use of computer predictions to tell us what is happening.

15

What is it?

My dictionary says that a greenhouse is a "temperate conservatory for growing plants" and that a conservatory is a "glasshouse". A greenhouse is a building made almost entirely of glass, which conserves heat and is used for growing plants. The leaves of the plants give the building a generally green appearance inside, and this, presumably, gives rise to its name. I can only assume that we do not call it a "glasshouse" these days, as this is the name given to an army prison. The name "conservatory" is now given to a glass building attached to the outside wall of a house, and for growing indoor plants or as a warm, relaxing extra room to the home. It is warm because glass allows the rays of the sun to pass through, but blocks the heat trying to escape.

The greenhouse heats up:
1. Short-wave rays from the sun pass through the glass.
2. The sun's rays inside are absorbed by plants, soil, etc.
3. Long-wave heat is radiated back.
4. Most heat is reflected by the glass. The temperature rises.
5. A small amount of heat escapes.

Perhaps you have noticed how hot it becomes in a parked car on a sunny day, even when the temperature outside is low. The sun's rays pass through the windscreen and are absorbed by the seats of the car. These rays warm the seats, which in turn radiate the heat back into the car. The heat reaches the windscreen and the windows but cannot escape through the glass. Thus the inside of the vehicle warms up. This is the way that greenhouses heat up.

The short-wave radiation of the sun passes through the glass walls and roof of the greenhouse and heats the floor, shelves, plant pots and everything else inside. These warmed objects pass back as long-wave radiation the heat they have gained, and as this tries to escape, most of it is blocked by the glass. So the heat

The atmosphere heats up:
1. Short-wave rays from the sun pass through the atmosphere.
2. The sun's rays are absorbed by plants, soil, buildings, etc.
3. Long-wave heat is radiated back.
4. Some heat is absorbed by gases and radiated back into the atmosphere.
5. Other heat escapes into space.

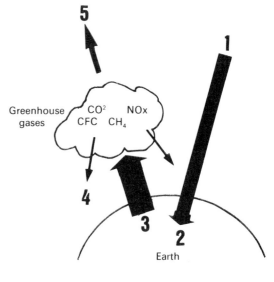

Greenhouse gases: CO_2 CFC NO_x CH_4

Earth

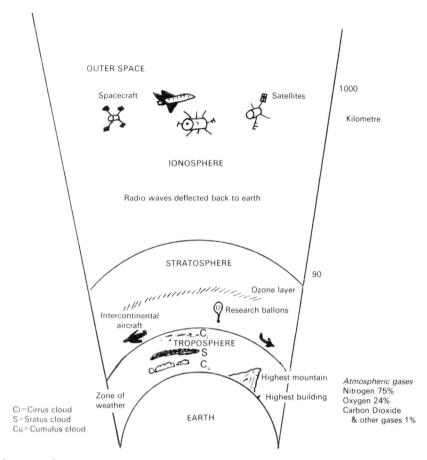

OUTER SPACE

Spacecraft

Satellites

1000

Kilometre

IONOSPHERE

Radio waves deflected back to earth

STRATOSPHERE

90

Ozone layer

Research ballons

Intercontinental
aircraft

C_i

TROPOSPHERE

S

C_u

Highest mountain

Highest building

Zone of
weather

Ci = Cirrus cloud
S = Sratus cloud
Cu = Cumulus cloud

EARTH

Atmospheric gases
Nitrogen 75%
Oxygen 24%
Carbon Dioxide
& other gases 1%

Above the earth: the atmosphere.

inside builds up. At night, with no extra sun entering, the heat eventually escapes, through the glass, or through the various gaps around the door and windows, or by being absorbed into the foundation of the building.

If the change in the climate of the earth is being explained by something called the Greenhouse Effect, then something similar must be happening on the earth to what happens in a real greenhouse. What is certain is that the earth has no roof of glass! It is surrounded with an atmosphere, made up of water vapour and many different gases of which nitrogen is the one to be found in the greatest quantity. For humans, the most important gas is oxygen. Carbon dioxide (CO_2) is another of the atmospheric gases. We take in oxygen and we breathe out carbon dioxide. In turn, this gas is absorbed by green plants, which return oxygen to the air.

The atmospheric gases all absorb heat, some more than others. The most important heat absorber is carbon dioxide. Other gases which take in a lot of heat are nitrous oxide (this is the one known as laughing gas), methane, ozone and chlorofluorocarbons (CFCs for short). These high absorbers are sometimes called the Greenhouse Gases.

The heat from the sun is absorbed by the earth and everything on it, such as forests, fields, plants, animals, rivers, lakes, buildings and roads. Everything which makes up our world absorbs some of the heat. As inside a greenhouse, these objects pass back the heat into the air around them. It is then that the greenhouse gases take in much of the heat and prevent it from escaping back into space. In a real greenhouse, glass *reflects* the heat. With the air, the gases *absorb* it. In reality, the term "Greenhouse Effect' is not therefore a correct scientific description, but the result is similar: the air warms up.

17

Warming the atmosphere

The Antarctic ice sheet. *Recently a block of ice about the size of Wales broke off and floated away.*

At an International Geographical Congress in Sydney, Australia, in early August 1988, the topic of the Greenhouse Effect dominated the discussion. Records kept since 1950 showed that there had been a rise in temperature on the Antarctic coast. The sea ice and ice sheets are particularly sensitive to global warming and extra ice melting may be detectable by satellite. The best way of detecting melting would be to record either the alterations in the height of the ice sheet or differences in the positions of the major outlet glaciers. This would require very accurate satellite-borne altimeters or a very long observation period. It is definite that one of the first indications of changes in global climate is to be discovered in the South Polar regions.

Without the atmosphere around the world, temperatures would be too low for ourselves and for the other creatures and plants which inhabit our planet. The atmosphere insulates the earth, preventing heat loss. But if too much heat is retained and not enough escapes, the air can become warmer and various changes in the earth's climate take place.

This can happen if some of the heat-absorbing gases increase in quantity, as, unfortunately, they appear to be doing. In particular, at this time in the history of the world, Carbon Dioxide (CO_2) is increasing enormously. The extra gas in the atmosphere then absorbs much more heat, preventing it from radiating back into the upper atmosphere and the space beyond.

The carbon cycle. *Carbon is in the air as gas, in the plants and animals as tissue, flesh and bone, and in the soil as a chemical.*

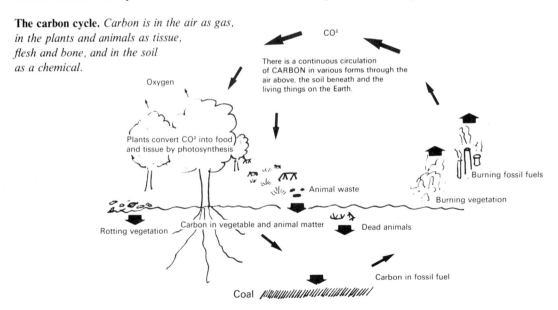

Of the seven warmest years on record, four have occurred in the 1980s.

Is it possible to predict by how much the temperature of the world will rise over the next tens of years? If scientists make certain assumptions – you could call them sensible guesses based on known facts – about how much of these gases will be released into the atmosphere, and then feed these facts into computers that can imitate the earth's behaviour, it is possible for "models" to be created of what might happen.

Several universities have research departments studying global climatic effects. Princeton University in the USA and the University of East Anglia in the UK are much involved. By studying climatic records and passing the information into computers, research workers at East Anglia have shown that there has been a steady increase in worldwide temperature over the past one hundred years. Different scientists have produced different theories, but most agree on a prediction that the Greenhouse Effect will lead to an increase in temperature of between 1.5°C and 4.5°C by the year 2030.

Water is much slower to heat up than land. As a result, the actual atmosphere of the earth will not warm up by this amount immediately, because the vast oceans, which makes up two-thirds of the world surface, will absorb about half of the increase. When they have warmed, perhaps fifty or more years later, the full effects will be felt.

The increase in temperature will not be spread evenly across the world. Some places will have a smaller increase in heat than others, if the theories of the scientists are correct. The real effect of the extra heat will also depend on the area to receive it. Pouring a kettle of boiling water into a bowl of cold water has a real effect. Pouring the same amount into a bowl of already hot water makes little difference. In a similar way, adding an extra three or four degrees of heat to the hot regions of the world may hardly affect the plants, animals or people there. The same increase at the polar ice caps could be devastating, as the ice would melt and the freed water add to the ocean levels.

A rise in sea level is only one of the results that may come from the Greenhouse Effect. Other serious problems may arise, as we shall see shortly.

Scenes such as this may become common in low-lying coastal areas all over the world as the sea-level rises. Eventually the floods will not recede and the former land will be submerged beneath the waves.

Sea levels

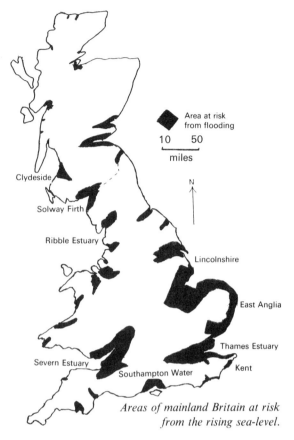

Areas of mainland Britain at risk from the rising sea-level.

Scientists estimate that almost 90 per cent of all the water on the surface of the earth is to be found in the ice of Antarctica. If the temperature rose enough for this vast frozen reservoir to melt completely, it is possible that the level of the sea would rise by 200 feet (60 metres) all around the world. The major cities of the world are on the coast. Imagine the effect on these of even a small rise of 1 metre. At the present time, the estimate is that the sea level is rising by at least 8 mm a year – a very small amount, yet enough to flood Central London, including the areas in which St Paul's Cathedral, Buckingham Palace and the Houses of Parliament are sited, by the later years of the twenty-first century. The Thames Barrier has been built to prevent flooding from high tides. It will need to be made higher to cope with the general sea-level rise.

Both the Antarctic (the South Polar region) and the Arctic (the North Polar region) are covered with enormous thicknesses of ice. The weight of the snow falling on them gradually squeezes the ice outwards, until it reaches a point where the surrounding oceans are warm enough to cause melting. Large pieces then float away as icebergs. Already there is evidence that the ice caps are beginning to shrink, as more ice melts owing to extra warmth in the air.

Glaciers are rivers of ice in the cold upper valleys of the mountains of the world. Like the ice sheets, they melt when they reach a point where the temperature is warm enough. The gradual warming of the atmosphere has led the ice of these glaciers to melt higher up their valleys. We say the glaciers are "retreating".

If the land is warming, then the water in the oceans is warming also, although at a slower rate. As water is heated, it expands and occupies more space. The effect is that the water level rises. Thus we have two causes for rising sea levels, both resulting from a rise in world temperature. Low-level countries such as Holland, and low-lying parts of other countries, such as around the Wash in Eastern England will have to build up their sea defences. Otherwise the land will have to be abandoned and people retreat to higher land, accepting the inevitability of the flooding. Even now in East Anglia exceptionally high tides and melting snow in the spring can bring about flooding.

Another problem will be the salination (salt pollution) of farming land and water supplies. When the flood waters have gone, the soil has become impregnated with sea salt and is unfit for growing many crops.

In the developed world there is enough technology, political goodwill and money to defend the land from spreading flood. Not so in

---

**Some sea-level cities –
find them in your atlas**

London   Belfast   Glasgow   Dublin
Cardiff   New York   San Francisco
Los Angeles   Melbourne   Sydney
Calcutta   Tokyo   Leningrad   Athens
Rome   Shanghai   Hong Kong
Rio de Janeiro   Buenos Aires

Located near Greenwich, the new Thames Barrier is designed to prevent the highest flooding that can be expected in normal times. It may well be unable to prevent flooding if sea-levels rise. The spaces between the piers shown in the picture are blocked by semi-circular barriers which rotate into position and which are hidden under the water when not in use.

The areas around the Wash, the Fenlands, are always at risk from flooding. Sluices such as this one at Denver, are positioned in the estuaries of all the rivers and drainage channels. Even now they have difficulty in coping with very high tides when melting snow is running from the land in early spring.

the developing countries. Bangladesh, for example, already suffers major flooding around the River Ganges delta. Here, too, is the site of the major Indian city of Calcutta. Tens of millions of people live there and around the coastal areas. The natural shape of the coast is a gently sweeping curve. This encourages the water of the Indian Ocean to build up in the Ganges estuary if the wind pushes at it from the south. Here, too, the depth of the sea is relatively shallow and this enables the power of the wind

to be more effective. The strongest winds occur in storm surges. They pile the water towards the Ganges, causing people to seek higher ground. Salt poisoning of the ground is intense.

A rise in the general level of the sea will only add to the regularity of these disasters until, eventually, the flooding remains as a permanent feature and the people are forced away as refugees. Extreme climatic events will no longer come at irregular intervals. They will be here to stay for several centuries at least.

Much of the coastal defences in Britain are little more than earth banks, too low to prevent the flooding associated with sea-level rise.

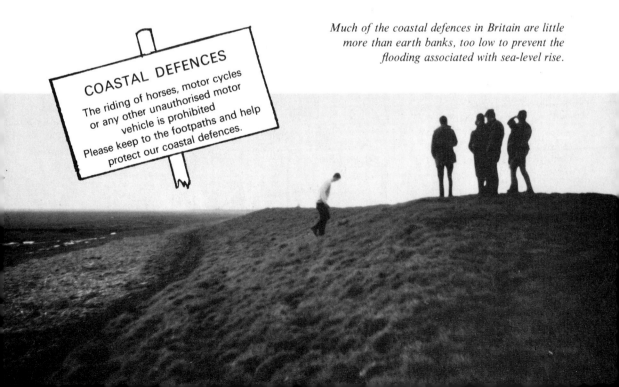

Crops

The increase in Carbon dioxide (CO_2) in the atmosphere and the change in climate resulting from the Greenhouse Effect (whether it be caused by more CO_2 or an increase in other gases) will lead to important changes in the growth of plants and, in particular, in the amount of food they yield.

The increase in CO_2

Extra CO_2 is known to increase crop yields. In commercial greenhouses three times the normal amount of CO_2 is pumped into the atmosphere. In this way, yields of tomatoes, for example, are increased tremendously. The CO_2 increases the photosynthesis effect, acting almost as a fertilizer on the plants. The amount of response depends on the type of plant. Agriculturalists recognize plants as C3 or C4 types: C3 plants show a great increase in growth with extra CO_2 and C4 plants show only a small response. 80 per cent of the world's food crops are C3, so this means that extra CO_2 will increase world food production. Unfortunately, many of the weeds which compete with our crops are C3 also and if C3 weeds are growing with C4 crops the chances of them choking the crops are obvious. In greenhouses weeds and pests are kept in check. In our fields this is not so easy and it is certainly more expensive of time and labour. Maize (corn) is an important C4 crop. Wheat is the main C3 one. An increase in CO_2 will help increase wheat yields; an increase in corn is less likely.

CO_2 also reduces the transpiration (losing water) of plants. Less water loss and enhanced photosynthesis increase the Water Use Efficiency of crops. Better WUE helps plant growth and higher yields.

But there has to be a snag! Increasing the CO_2 may increase the crop yield, but at the same time it reduces the nutrient (food) value of the plant. Experiments with caterpillars on leaves which have had extra CO_2 applied show that the caterpillars then need to eat much more of the plant in order to survive. The caterpillars therefore cause greater crop damage and this decreases the yield, unless better pest control is used.

C3 and C4 crops

Common C4 crops are Maize, Sorghum, Millet, Cane Sugar.

Common C3 crops are Wheat, Rice, Soya Bean, Beet Sugar, Bananas, Grapes, Oats, Rye, Barley, Cassava, Potato, Sweet Potato.

Doubling CO_2 increases growth and yields by 10 per cent to 50 per cent for C3 crops and by 0 per cent to 10 per cent for C4 crops.

Experiments in laboratories cannot imitate the real-life situation. Here an experiment in the Chesapeake Bay area of the eastern USA is taking place "in the field". An array of gas pipes is releasing 400 tonnes of CO_2 a day onto the vegetation – double the amount they normally receive – to see how they adjust to increased CO_2 in the atmosphere.

Sweet potato, a very important crop in many underdeveloped countries, undergoing high CO_2 levels produces much less protein, so that there is a need for more of the crop to produce the same amount of nutrition.

The change in climate

Changes in temperature and in the amount of rain and snowfall affect vegetation and thus food crops. If one result of the Greenhouse Effect is to cause the rain belts to move northwards in the northern hemisphere, then the main grain-growing areas of the world will move northwards, too. Above all other food-producing areas, the grain-growing regions of North

Warmer winters in the grain-growing areas.

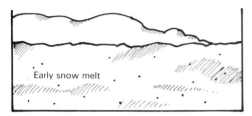

Early snow melt

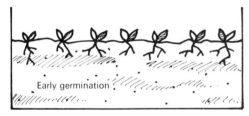

Early germination

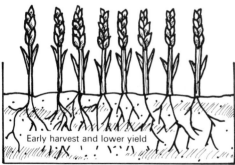

Early harvest and lower yield

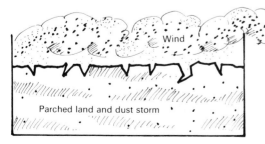

Wind

Parched land and dust storm

Warmer winters in the Grain Growing Areas

A "grain mountain" being tested for heat and damp in one of the European Common Market stores.

High yields of grain in America and Europe have led to surpluses until now. The excess has been stored and used in times of shortages in other parts of the world. Extra CO_2 may increase yields further, but a move northwards of the grain-growing areas may lead to an even greater reduction.

America are the "safety net" against world food shortages. Anything which affects them adversely is very serious indeed. The effect of warmer air on North America's winter snowfall has the following results:

1 Warmer winters cause the snow cover to melt earlier in the spring. The soil becomes moist earlier and the autumn-sown grain germinates into life that much sooner.

2 The air remains warmer, and this speeds up plant growth. The summer season lasts longer and more water evaporates from the soil. The crops mature more quickly and give a lower yield of grain.

3 The crops are harvested earlier and, unless precautions are taken to prevent the soil cover being left bare, the dried-out soil will blow about as dust, recreating the Dust Bowl conditions discussed earlier (pages 6–7).

It would be wrong to conclude that, if the main crop areas moved further north, this would simply mean a changed location for the grain harvest. Other areas of North America may not be as suitable for grain production, despite the suitable climate now in the new area. Soil and other conditions have to be right, as well as the weather.

A final question which we cannot answer until the actual events occur is whether the positive advantages to crop growth of extra CO_2 in the air will be enough to override the disadvantages of climatic alteration.

Clouds

One of the pleasant things about flying is that the sun is always shining once the aircraft has risen above the clouds. The clouds stretch in all directions, looking solid enough from above, and bright white from the reflection of the sun upon them. Having come through the cloud, the plane has proved, if proof were needed, that the clouds are actually far from solid. They are made up of water droplets held aloft in the atmosphere at the level where the temperature is low enough for the water to condense out of the air. This is known as the Condensation Level. From below, the clouds look black, grey or white depending on the amount of light which passes through them. On a very wet day, much of the sun's light and heat is blocked by the clouds, although an enormous amount must reach the ground, otherwise it would be as dark as night and much colder than usual.

The clouds at the lowest level in the sky are massive and rounded, shaped rather like a cauliflower. They are the most common kind of clouds we see and are above us on most days. Their name is Cumulus. Higher up on dull days there are sheets of flat grey cloud. These are the Stratus clouds and, of all the cloud types in the sky, they reflect the most sunlight. Very high in the atmosphere, wispy clouds called Cirrus rarely block out the sunlight completely. Even on a bright sunny day they are present in a minor way. They block the earth's heat trying to escape into space, but not the sunlight coming in.

The Stratus clouds prevent heat reaching the earth and so decrease the Greenhouse Effect. The Cirrus clouds trap the heat of the earth and thus increase the Greenhouse Effect. The low-level Cumulus are usually so broken up that they make little contribution to the temperature balance. Meteorologists admit that they do not know precisely what the overall effect of clouds is on the Greenhouse Effect, or what will happen with global warming. They believe that the middle-level clouds will decrease, permitting more sunshine onto the earth. They also think that the high-level clouds will rise even higher, so that they will become a more effective block to escaping heat. As to the low-level clouds, they believe that they will increase. On balance, it is thought that as the increase in greenhouse gases warms the earth, this will increase the cloud cover, especially of high-level Cirrus, and so magnify the greenhouse warming effect. However, clouds remain a source of major uncertainty as scientists try to assess what is really

Watch cloud formations. See if you can identify the cloud types. Observe carefully for a little while and see how the clouds are always changing shape.

CIRRUS: UPPER AIR CLOUD

STRATUS: MIDDLE LEVEL CLOUD

CUMULUS: LOW LEVEL CLOUD

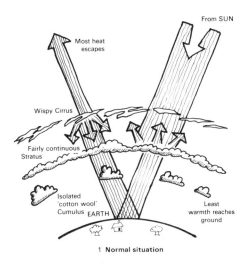

1 Normal situation

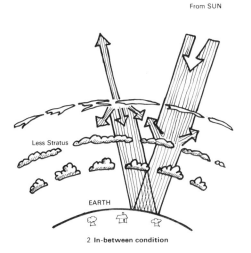

2 In-between condition

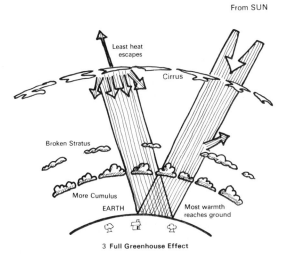

3 Full Greenhouse Effect

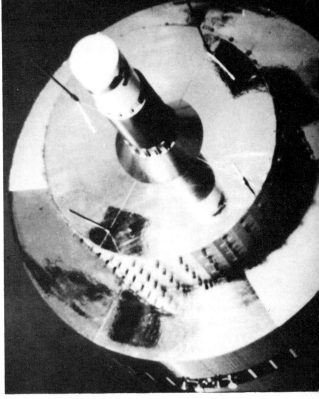

Meteostat satellite in space. *It transmits information about the clouds and the weather to earth.*

going to happen to our climate.

Satellites now enable us to observe and record the clouds, especially those at high levels, and it is hoped that, as a result, more will be understood about them. Direct observation is very important to confirm the remote investigations by satellite. Clouds which have been observed by a satellite have been directly investigated by high-flying aircraft equipped to measure all of their aspects, particularly the amount of the sun's radiation passing through them and the amount of the earth's heat radiation passing back into space. The ice particles which make up the very high clouds are electronically photographed, and the images obtained are studied in the laboratory back on earth. So far the conclusion confirms the difficulty of assessing the impact of cloud in judging the Greenhouse Effect.

Cloud prediction as CO_2 increases in the atmosphere. *Normally, stratus clouds reflect heat back into space; and cirrus clouds allow much of the earth's heat to escape. With the full Greenhouse Effect, broken stratus clouds reflect only a small amount of the sun's rays. Cirrus clouds at a greater height allow only a little heat to escape.*

Carbon dioxide – fossil fuel

The Drax coal-burning power station near Selby in Yorkshire can produce 4,000 Megawatts of electricity.

At the present time it is estimated that at least half of the warming due to the Greenhouse Effect is caused by CO_2, although it is possible that other gases will become more important in the future. Where does CO_2 come from? The greatest amount is produced when vegetable matter is burnt. In particular, CO_2 is a product of burning fossil fuels (coal, oil, gas) in industrial processes and in electric power stations. For example, the Central Electricity Generating Board (CEGB) states that a coal-burning power station which produces 2,000 MW of electricity emits 10 million tonnes of CO_2 each year. In 1987, coal-fired power stations operated by the CEGB were capable of producing a total of 30,537 MW of electricity, although it was unlikely that all of this capacity would have been used

at any one time. You can calculate that all these generating stations working together would be capable of pushing 150 million tonnes of CO_2 into the atmosphere every year. And this is only for England and Wales. Imagine the amount of CO_2 produced when the power stations of Germany, Russia, China, the USA and all of the rest of the power-making nations are taken into account. Our minds cannot appreciate just how much this is, except to know that it is far, far too much.

Every other industrial process which burns coal, oil, gas, wood or any other vegetable matter adds to this staggering load of CO_2 emission. A good example (perhaps I should say a "bad example") is a British Coal works in South Wales, where they are producing "smokeless" coal. Amongst the fumes which are blighting the lives of the local residents is CO_2. Yet the whole idea of smokeless fuel is to prevent domestic fires sending noxious smoke into the air.

Bricks and china are made by heating clay, turning it from a weak substance into a hard material. Electrically heated kilns have replaced most of those fired by burning fossil fuel, but it was only a short time ago that "The Potteries" were belching out smoke into the air of Staffordshire. Brickworks are identified easily by their many chimneys close together, even nowadays emitting smoke into the atmosphere. If you travel by car on the M1, or by train, near to the new town of Milton Keynes, it is obvious that the brickmakers of Bletchley are not far away. Even if the chimneys themselves are out of sight,

Material flows for 2000MW coal-fired station

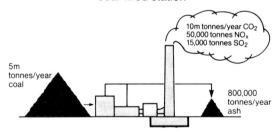

On the basis of this CEGB diagram, the Drax power station burns 10 million tonnes of coal and ejects 20 million tonnes of CO_2 every year. The Flue Gas Desulphurization plant being installed at Drax will deal with the sulphur dioxide which causes Acid Rain, but will not reduce the CO_2 or the NOx.

Oil refinery, Ellesmere Port. *Oil, another fossil fuel, needs to be refined into petrol or diesel before it can be used. Both refining and use in engines give rise to emissions of greenhouse gases, including CO_2.*

the "sweet" smell in the air is clear evidence of the brickworks.

Another source of CO_2 is the burning of rubbish. As sites for tipping household and industrial rubbish have become fewer, many UK councils have established incineration plants. Cement manufacturers, the steel and metal industries and many others add to the CO_2 pushed out and up.

From the point of view of overloading the atmosphere with CO_2, it does not matter where the gas is made; what matters is the total amount produced and sent into the earth's atmosphere. To be fair to industry, much has been done to reduce emissions of CO_2 and other gases – but much more action is still needed.

Hospital chimney, Walsall, in the West Midlands.
Hospitals have to dispose of much unpleasant unwanted material. By far the best way to destroy infectious material is to burn it. The tall chimneys of every modern hospital complex can be seen, every day, sending out noxious smoke into the air. Such chimneys have "Crown (government) immunity", which means that they do not have to obey local regulations on smoke control.

Carbon dioxide – burning vegetation

Burning vegetation anywhere produces CO_2. Bonfires in our gardens or on November 5th, cooking fires in Africa and the Far East, funeral pyres – can you think of other deliberate burning of vegetation? Certainly the most important is the burning of the tropical rainforest, which is estimated to produce one-quarter of the increase in the world's CO_2. The "slash and burn" method of agriculture, and the burning away of vast tracts of rainforest to provide grazing ground for cattle destined to become the meat in hamburgers, are described in the book *Disappearing Rainforest* in this series. Later we shall see that the cattle which replace the forest are themselves responsible for adding to the Greenhouse Effect. Estimates vary as to the loss of rainforest every year, but a recent reliable source put it at a minimum of 20 million hectares.

Land clearance in Borneo by burning the rainforest.

Burning the Amazon Forest

In September 1988, Herbert Girardet travelled across the Amazon rainforest. He described what he saw:

"The savannah country below us was on fire. From horizon to horizon columns of smoke were rising thousands of feet into the air. Hills were hardly visible. Wherever we landed on our seven hour flight, grass and small trees were on fire.

I could not believe the scale of what was happening. As we took off again from Conceição for the last leg of our flight to Redeneçao, the smoke became so thick that the pilot lost his way. Some of the passengers had tears running down their cheeks. The acrid smoke, the feeling of helplessness and outrage in the face of so much destruction left us gasping.

Eventually the pilot did find the air-strip at Redeneçao. As we arrived in this frontier town of gold diggers and cattle ranchers—only 50 years ago the location of an Indian jungle village – the smoke was so thick that one could not see across the street.
(Eyewitness report in the *Daily Telegraph*, September 29th, 1988)

In Brazil, the right to own land can only be given when half of the forest has been cleared away. A simple change in the law of that country might encourage people to retain, and not destroy, the rainforest.

One famous fast food chain sells 145 hamburgers every second; has sold over 50,000,000,000; serves 19,000,000 customers daily. AND then there are all of the other burger-selling restaurants!

At Carajas in the eastern Amazon forest of Brazil, two new charcoal-making furnaces have been opened. Twenty such furnaces are planned, to provide fuel for making pig iron. One such iron works opened in Maraba in March 1988, using money provided as a loan by the World Bank and the European Economic Community. When the twenty charcoal-making furnaces are working, it is thought that 450 square miles of forest will need to be cleared for fuel every year. Work out what 450 square miles would mean to you in your locality. Opposition to the plans is intense, and second thoughts on behalf of the world financiers may mean a stop to further loans for the projects.

The burning of straw and other plant stubble in Britain's fields is always a controversial issue during late July and August. Farmers who do it claim that they need to clean their fields of weeds, insects and vermin, as well as ridding themselves of unwanted vegetation. Yet elsewhere in our country there is a great demand for straw for animal bedding. It is certain that the thick clouds of smoke are dangerous to passing traffic, obnoxious to local people, a menace to hedges and wildlife and another source of CO_2. The burning of heather and bracken is another land management practice which pollutes the air with burning vegetation.

Accidental forest fires add to the problem. 1988 saw the burning of large areas of forest in North America associated with the dry conditions described earlier. In this way, the drought-causing Greenhouse Effect, itself caused by excess CO_2, was then causing even more of the gas to be sent into the atmosphere. We call this a "vicious circle".

Volcanoes

Other natural processes add to the CO_2 in the atmosphere. The most obvious are volcanoes. Even where volcanoes are not in spectacular eruption, many of them are showing signs of activity all of the while. If you visit or fly over Sicily, at the foot of Italy, you will see the smoke rising from Mount Etna and the surrounding "fumeroles", places where very small puffs of smoke occur most of the time. It is the same with all sites in the volcanic areas of the world. Although the headlines of our newspapers report only the violent eruptions which cause death and destruction, volcanic activity, over which we have no control, is always going on.

Smoky gloom is brought to the English Midlands as a farmer burns off straw stubble on a bright sunny day. The nearby M54 was badly affected.

Carbon dioxide – facts and figures

But what about ourselves? Our bodies exist by inhaling oxygen and exhaling CO_2. So, with every breath we take, we ourselves are adding CO_2 to the air! Not that we can do much about this – we can hardly stop breathing as a practical solution! However, the world population is increasing at an alarming rate, and it is essential to control this, not just because of the impact more people might have on the Greenhouse Effect, but for other reasons, too, which may be even more important.

The greatest growth in population is taking place in the developing world. More people means that more land is cleared for agriculture and more fuelwood is used for heat and cooking. Both these activities involve increased burning and thus create extra CO_2. As the governments of the less developed nations seek to improve the living standards of their peoples, it seems inevitable that they increase power production by using fossil fuels. This may have to be taken into account by developed countries, who could make up for the extra CO_2 by reducing even further their own emissions of greenhouse gases.

Trees, and other plants, "breathe" as well, taking in oxygen and giving out CO_2. They also carry out a reverse process (taking in CO_2 and giving out oxygen) which is essential to the control of CO_2 in the atmosphere as we shall see shortly.

Nonetheless, whatever control methods might be employed, the task of preventing further increases of CO_2 in the atmosphere is stupendous. Levels of the gas have been increasing for 200 years, since the start of using steam power at the beginning of the Industrial Revolution. It has been estimated that in 1850 there were 265 parts of CO_2 in every million parts of the atmosphere. In the 1980s it is about 340 ppm (parts per million) and the fear is that, by the year 2050, it could be over 600 ppm unless a drastic reduction takes place in the use, and burning methods, of fossil fuel and in the devastation of the rainforests.

Such facts and figures come from long-term investigations into the quality of the atmosphere. For instance, the USA has a scientific programme called the Geophysical Monitoring for Climatic Change. One of its bases is at the

Early industry belched out smoke with no pollution control at all. Here, 100 years ago, the Attercliffe Steelworks in Sheffield were typical of the landscape of manufacturing towns.

In 1987 world population was estimated to be 5000 million. More people demand more energy, food and goods. As things are now, this means more atmospheric pollution.

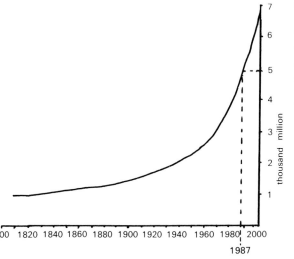

years, to give one of the scientific world's longest continuous records of CO_2 in the atmosphere. Statistics confirm that the amount of the gas in the atmosphere has increased by over a quarter during the last 140 years.

If the forecast for the future is correct, it will mean that a doubling of CO_2 will have taken place in the 200 years since 1850. This cannot be an acceptable situation. Imagine if a single country somewhere announced that it intended to add that amount of CO_2 to the air. There would be a world outcry and intervention by the United Nations. Because the extra gas is coming from innumerable sources worldwide, less attention is paid to the matter.

top of Mauna Loa, an extinct volcano in Hawaii. Over 4000 metres high, it is mostly above the clouds in bright sunlight and supposedly clean air. Special instruments with complicated names "taste" the air. Nephelometers, spectrometers and transmissometers sort out the gases. The site has been researching the air for very many

*Carbon dioxide emissions into the atmosphere have increased the CO_2 content of the air above Mauna Loa dramatically in the last 30 years. (*Source: Environment Canada)

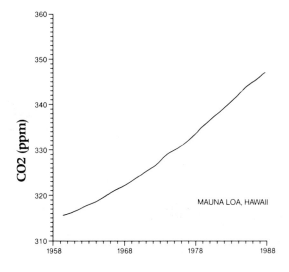

Burning wood is the basic source of heat for the majority of people in the underdeveloped world. Collecting fuelwood is a daily task. Better stoves and other conservation methods will, hopefully, restrict the destruction of trees, while still allowing improved living standards.

Carbon dioxide and drinks

The bubbles in "pop", beer, cider and other drinks are CO_2 gas. Modern public houses use CO_2 from gas cylinders in their cellars, to force the beer up to the bar. All this CO_2 eventually escapes into the atmosphere

Carbon dioxide – technological control

If most of the CO_2 comes from the burning of fossil fuels, then this is where controls must be applied. At the present time, countries of the developed world are the ones using fossil fuels to power their industries and the modern technology of domestic living, with its cars, TVs, washing machines, cookers and the like. This is particularly true in heavily populated urban areas and less true in rural regions. The developing world is striving to catch up with what appears to be the desirable affluence of the West. Those countries want to establish their own industries and electric power, and this will require the burning of fossil fuels or of vegetable matter directly, particularly wood. The chances of reducing emissions of CO_2 without world agreement are remote. Extra CO_2-saving efforts will have to be made in the already developed regions to balance the inevitable industrial growth and the increase in CO_2 emission in the developing world.

The amount of CO_2 produced by burning fossil fuels can be reduced if we:

(1) *use less energy*. This can be brought about by the efficient use of what is already available. Energy conservation methods that will reduce demand include insulating roofs, walls and windows, draughtproofing buildings, making a slight alteration in thermostat settings for central heating and air conditioning, and using off-peak electricity. Increasing standards of living seem to demand more electricity. By not wasting electric power we can prevent the need for more production.

(2) *use better technology*. It is possible to burn fossil fuel more efficiently so as to make better use of the heat available and, at the same time, in the process, destroy most of the waste gases. The CEGB's coal-fired power stations have a Thermal Efficiency of between 17 per cent and 36 per cent. This means that, for every 100 tonnes of coal burnt, only between 17 and 36 tonnes actually make the electricity; the rest goes towards making heat, which is wasted. The larger power stations are the most efficient, but even at 36 per cent efficiency almost two-thirds of the power of the coal used is lost. Even when the power has been made, some of it is lost in

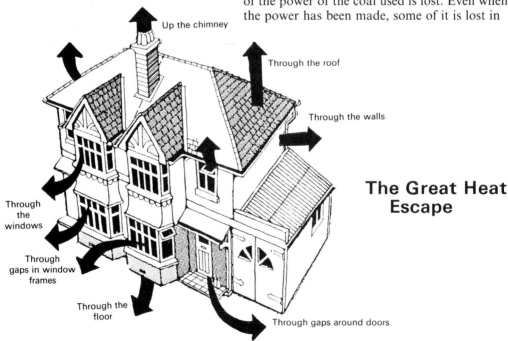

Up the chimney

Through the roof

Through the walls

Through the windows

Through gaps in window frames

Through the floor

Through gaps around doors

The Great Heat Escape

Energy conservation is the most important single way by which the burning of fossil fuel could be reduced. Insulation of our homes is a first step. It also results in reduced fuel bills.

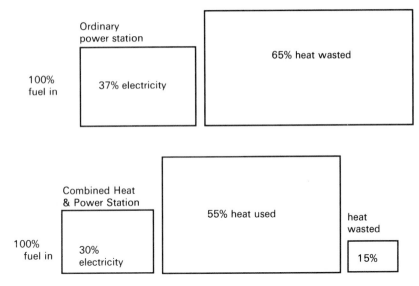

Ordinary power station

100% fuel in

37% electricity

65% heat wasted

Combined Heat & Power Station

100% fuel in

30% electricity

55% heat used

heat wasted

15%

A comparison between wasted heat at an ordinary power station and at one designed for Combined Heat and Power. Reducing waste saves energy. By saving energy, you reduce the need for electricity, and thus less atmospheric pollution is created.

being transmitted to the user along the grid lines, and even more is wasted through poor equipment in home, factory, office, and shop. It is possible for the greenhouse gases produced by inefficient furnaces to be reduced by cleaning equipment in the chimneys.

(3) *introduce new technology to use the wasted power.* The UK makes little use of Combined Heat and Power techniques. With CHP the

unwanted steam and hot fumes escaping into the air are used to provide useful heat in buildings nearby. Hot water for central heating and ordinary family use can be carried to the consumer from the power station. Vasteras, a town of over 100,000 people in Sweden, has a 90 per cent efficient CHP plant. In other words, only 10 per cent of the heat is lost. The exhaust gases from the coal-fired power station are reduced to a minimum and, because hot water is supplied to all the buildings in the town, no other furnaces are necessary and so no other CO_2 is produced.

(4) *use alternative methods for producing power.* Hydro-electricity is very efficient and totally free of exhaust. Greater use could be made of the direct power of falling water, specially in small units, with the use of pumped storage adding to the overall efficiency (see *Energy, Power Sources and Electricity* in this series). The alternative energy sources of wind, tide, solar, geothermal, biogas, and wave power remain to be exploited everywhere.

If these methods of reducing CO_2 emissions were sincerely and urgently applied, with international agreement, it is estimated that a reduction of 60 per cent could be achieved.

District hot water supply. *Hot-water pipes being laid in Denmark, to use the waste heat from a CHP station for a central heating and hot water supply to houses and factories.*

33

Carbon dioxide – control by plants and . . .

Do you know what phyto-plankton is? We rarely see it, for it is minute plant life floating on the oceans. It is important because it provides food for small creatures which in turn are the food for fish. It is equally important because it acts with CO_2 like any other vegetation.

I mentioned on page 30 that, whereas human breathing takes in oxygen and gives out CO_2, plants take in CO_2 and release oxygen. They do this in a process called photosynthesis. Of the two processes, photosynthesis is by far the more important, as far as the amounts of gases are concerned. Plankton is very valuable in exchanging CO_2 for oxygen, but it is mainly beyond our control, except that we must prevent pollution of our seas and oceans as far as we can. Trees are the major recyclers of CO_2, and we can take action about trees. Planting new trees is valuable, as every new tree planted is a help to reduce the CO_2.

The problem of excess CO_2 is a worldwide one, and so worldwide thought needs to be given to its solution. As mentioned earlier some 20 million hectares, about 77,000 square miles, of rainforest are being destroyed every year – over 200 square miles each day. Why not stop this? The politics and the economics of the problem are discussed later, but the system described on page 28, where land in Brazil can be owned only after most of it has been cleared of forest, is but one example of the many foolishnesses that occur.

You could take the present rate of destruction of rainforest and say that the amount of forest being lost in two days would have been able to neutralize the CO_2 from a coal-burning 1000MW power station. This would be a very simplified view, and the problem is much more difficult to resolve than my numbers suggest. But they do highlight the damage we are doing to our world environment. When did the front page of your daily newspaper mention any of this? We must recognize the importance of the control of CO_2 by plants.

Leaves – the vital recyclers of CO_2. *Do you recognize the sort of tree?*

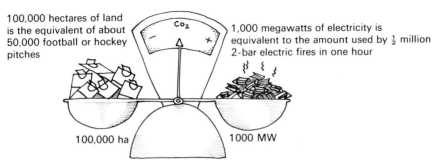

100,000 hectares of land is the equivalent of about 50,000 football or hockey pitches

CO_2

1,000 megawatts of electricity is equivalent to the amount used by $\frac{1}{2}$ million 2-bar electric fires in one hour

100,000 ha

1000 MW

It has been estimated that the CO_2 emitted from a 1,000 Megawatt coal-burning power station would need 100,000 hectares of woodland (almost 400 square miles) to "devour" it completely. Maybe not a practical possibility. Far better to prevent the enormous destruction of rainforest.

... and another way to reduce CO_2 : use nuclear energy

Another answer to the problem of energy production polluting the air with CO_2 and other exhaust gases is to USE NUCLEAR POWER INSTEAD OF FOSSIL FUEL. Nuclear power stations do not send out any exhaust fumes. No CO_2 or other pollutant gases are emitted from them.

However, there is the fear of the spread of radioactivity. Accidents do occur, as we all know from incidents at Sellafield (UK), Three Mile Island (USA) and Chernobyl (USSR). Such accidents must influence our thinking, as must the suspicions that illnesses like leukaemia are more likely to affect people living near to nuclear installations, particularly those dealing with radioactive waste. The problem of atomic waste disposal is probably even more important for us to consider than the escape of radioactive clouds.

Dealing with a nuclear power station when its useful life is over is not easy. It takes tens of years to close one down and then parts of it have to remain sealed far into the future.

The atomic process of power production through safe nuclear FUSION, rather than nuclear FISSION, may be the answer one day. At the moment, nuclear power is feared by many people, despite its many advantages – not least of which is its ability to produce electricity without any CO_2.

Heysham, near Morecombe, Nuclear Power Station. *No chimneys, no smoke, no CO_2 or other gases.*

France produces most of its electricity from atomic power and claims that it saves making 120 million tonnes of CO_2 every year.

Coccolithophores

You may well ask, "What on earth are coccolithophores?" A better question would have been, "What in the oceans are coccolithophores?", for they are part of the microscopic phyto-plankton mentioned previously (page 34). Although they are plants, they have the interesting ability to form skeletons of calcium carbonate. These skeletons take on wonderful shapes, some like plates, others like snowflakes, car tyres, starfish, and playing cards spread out into a circle. They are incredibly small, with fantastic numbers like 100 million of them to every litre of water. It is only under powerful microscopes that they are visible as individual objects. Together, to the naked eye, they make the water take on a milky colour and appear as giant floating patches in the oceans, known as "blooms".

These blooms cover vast areas of over 400,000 square kilometres. These days they are observed from satellites, which have recorded two main swirling patches in the North Atlantic – one to the north-west of Scotland, near the Rockall Bank, and the other just south of Iceland. Each is about the size of Britain, and historical records reveal that they appear every summer.

When coccolithophores die, their minute skeletons fall to the bottom of the ocean to form a muddy deposit known as "ooze". Eventually, over millions of years, these oozes harden to become rock, which makes the upland areas when sea levels change. Chalk is made in this way, so that when you see cliffs such as the white cliffs of Dover, you are looking at innumerable coccolithophores and the remains of other shelly creatures.

Why mention them at all in a book about climate? Like all plants, these algae absorb CO_2, in their case to make their protective shells. They also reflect sunlight back into space. Both of these actions help to prevent the rising temperatures associated with the Greenhouse Effect. But coccolithophores do something else which affects the weather. They emit sulphur chemicals into the air. On-the-spot observations by marine research ships, such as the British *Challenger*, have led scientists to conclude that these blooms are a major contributor to the formation of acid rain. Claims have been made that as much as half the sulphurous chemicals and gases which react with rain to make it acidic come from such biological sources, where the country receiving the rain has a long coastline. Norway would be a good example of this. Acid rain falling on Scandinavia is mostly blamed on the sulphur emissions from power stations in the British Isles. If, in fact, half of the sulphur in the atmosphere does come from algae bloom, this would remove

Phyto-plankton magnified over 1,000 times reveals a variety of shapes and sizes.

36

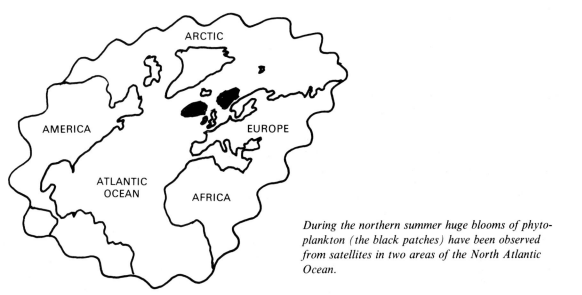

During the northern summer huge blooms of phyto-plankton (the black patches) have been observed from satellites in two areas of the North Atlantic Ocean.

some of that blame. Proving it will be extremely difficult. Much more research, extending over many years, will be necessary to substantiate the acid rain claim and, perhaps even more important, to find out exactly how the coccolitho-phores alter the climate and what their "biogenic" effect is, as it is called by scientists. Much of this research will be carried out by "remote sensing", the term given to obtaining information on the environment by satellites high above the surface of the earth. The main worry of those involved is whether or not money will be available to continue this important research. Satellites are very, very expensive to make, launch and use.

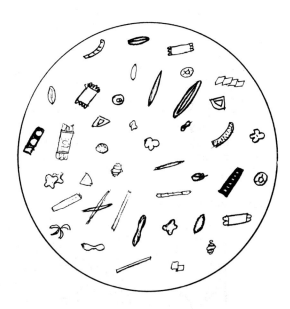

Under the microscope, a cross-section of a piece of chalk shows that it contains the remains of plankton.

Pollution

Oil spills, chemical waste, sewage all affect the plankton. Small creatures, the ZOO-PLANKTON, feed on PHYTOPLANKTON. Fish and other creatures eat the ZOOPLANKTON. Ocean pollution harms the whole food chain.

Obtaining the facts

Facts about the climate have to be gathered from land, sea and sky before future conditions can be considered and before information can be fed into computers. All around the world and all around the clock, meteorologists are analysing past statistics, recording the present and forecasting the future.

From the sky (*top left*). *The C-130 aircraft of the Meteorological Research Flight collects information from high in the atmosphere.*

From the South Polar region (*bottom left*). *The launch of an ozone-sonde in Antarctica. The radar on the left will track the instrument, which sends back information on the ozone concentration.*

From the sea (*top right*). *Weather instruments on buoys send back readings automatically. An engineer from HMS* Hecla *checks the equipment in the Atlantic.*

A national data buoy off the south-west coast of Britain (bottom right). Its exact location is 48° 40′ N, 09° 00′ W.

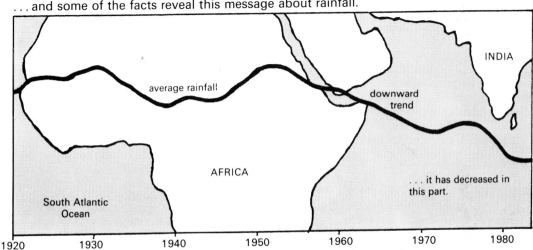

1920 1930 1940 1950 1960 1970 1980

North Atlantic
Ocean

U.S.S.R.

UK

Upward Trend

EUROPE
average rainfall

As much as the
rainfall has increased
in this part of the world . . .

Mediterranean Sea

. . . and some of the facts reveal this message about rainfall.

average rainfall

INDIA

downward
trend

AFRICA

. . . it has decreased in
this part.

South Atlantic
Ocean

1920 1930 1940 1950 1960 1970 1980

THE MIRROR IMAGE. *As average rainfall in the African part of the world between 5°N and 30°N is* **Down**, *in the European part between 35°N and 70°N it is* **up**. (Source: University of East Anglia)

From the land. *A complete radio-sonde rig is launched. The recording instruments are in the triangular bag which hangs from the balloon. Signals are received back at base.*

From schools. *Every school can have a weather station. Accurate recording of the facts is something all can do. Here pupils are using a rain gauge and thermometers in the Stephenson screen (the white box).*

CFCs: Chlorofluorocarbons

Some gadgets and inventions seem to have been with us always, but this is obviously not the case. When, for instance, were the spectacle lenses introduced which go darker as the sun shines on them? What about the covering on pots and pans which makes them non-stick for cooking? When did compact discs, car radios and walkman radio-cassette players come along? Before or after you were born? Five, ten, twenty, or more years ago? (See foot of this page.) These are examples of scientific discoveries and inventions now applied to day-to-day living.

Another is the aerosol spray. Aerosols are actually very small droplets of liquid, and the spray-can is the invention used to project them out as a fine mist. Spray-on paint for repairing scratches on cars, insect-killing chemicals for dealing with flies, aphids and other creatures, deodorant liquids to prevent body odours, lacquer for keeping hair in place, and many more are now common in our daily lives. New can construction techniques, coupled with a method of using special gases to propel the liquid into space at the press of a button, made aerosol sprays a practical possibility.

Polychromatic material for anti-glare sunglasses, Teflon coating for non-stick pans, micro-silicon chips for radios and cassettes, and CFCs as propellant gases, have made their mark in our daily lives. The snag is always that a new technique may have unwanted side-effects. In the case of CFCs, there is a very serious side-effect. It is now quite obvious that CFCs in the atmosphere absorb heat from the earth and prevent it escaping. Like CO_2, CFCs are greenhouse gases. Much publicity has been given to the problem of CFCs being emitted by aerosol sprays. But CFCs are widely used elsewhere. They can be used in the foaming process to make expanded plastic material for seats and cushions. One type of this plastic material can be used to make packaging such as egg boxes, beakers, sandwich boxes and food trays. CFCs are also used as solvents for removing the chemicals used in constructing the electrical equipment in computers. Another use for CFCs is in the cooling processes for refrigerators and air-conditioning equipment. One of the aims of the Chinese government is to provide a refrigerator for every family by the year 2000. With a thousand million people in China, that makes an enormous demand for the production of CFC gases. If countries with huge populations like India and China do not join the world agreement to restrict the use of CFCs, it will make the efforts of other countries almost pointless.

CFCs have many advantages, the most important being that they do not chemically affect or contaminate anything which they touch. They are said to be chemically inert. They just do their job and pass on. Unfortunately, we now realize that they pass on into the air to become a further cause of the Greenhouse Effect – and something else perhaps even more important, as we shall see.

Another gas, closely related to CFCs, is Halon. This is used in fire extinguishers, to turn the flame-killing chemicals into a foam. Halon, too, is a greenhouse gas.

Spraying from an aerosol can is a common domestic practice nowadays. From most aerosols CFC gases are released.

Car radios	over 60 years ago
Walkman sets	about 10 years ago
Compact discs	about 5 years ago
Non-stick Teflon	about 25 years ago
Photochromatic lenses	about 20 years ago

Foam packaging is made by using CFC gases as a foaming agent. When these articles are crushed, a little CFC is released.

A typical collection of aerosol cans from one household. Not all of them contain CFC gases. Soon it is hoped that non-CFC cans will be labelled.

Chlorofluorocarbons (CFCs)

CFCs are made up of different proportions of carbon, hydrogen, chlorine and fluorine atoms. Two major types of CFCs, CFC 11 and CFC 12, are used in foams, aerosols and chillers for refrigeration. They are inert, non-toxic and non-flammable; that is, they are chemically inactive, they do not poison and do not catch fire. But they do destroy ozone and absorb heat. In fact, they are 10,000 times more effective as a greenhouse gas than CO_2. The search for ozone-"friendly" gases, which can do the same job as CFCs, is going on. It is generally accepted that CFC12 could be replaced fairly easily. Evidently it is more difficult to find a substitute for CFC11. The basic idea is to replace the chlorine content of the gases with hydrogen. These gases will be known as HFCs, with HFC 134a the first to go into production to take the place of CFC12. Unfortunately, the cost is going to be treble that of the CFC it will replace. The numbers of the gases reflect the ratios of the different chemicals in the gas.

Ozone

Jekyll and Hyde – do these names mean anything to you? They are the names of R. L. Stevenson's famous fictional character, Dr Jekyll and his evil other half into which he could change. As Dr Jekyll, he was good; as Hyde he was bad. The gas ozone is rather like Jekyll and Hyde. It provides a vital protective shield around the earth, but it can also be a danger to us.

Ozone is oxygen with another atom added. Chemically, oxygen is made up of two atoms – O_2. Ozone consists of three atoms – O_3. Oxygen is stable; it does not change easily. However, given certain circumstances, it will become ozone. This change happens where sunlight acts upon the nitrogen oxides in vehicle exhaust. In what is called a photo-chemical process, free atoms of oxygen extracted from the car fumes by the sun's action link with the oxygen in the atmosphere and ozone is formed. In places such as Los Angeles, the basin-like shape of the land causes the photo-chemical process to produce health-hazardous smog. At higher altitudes, the ozone which is created has an adverse effect on trees. In the Black Forest of Germany, for instance, acres of wooded hills are littered with dead and dying trees, a phenomenon which is part of Acid Rain pollution.

High in the atmosphere, several miles above the earth's surface, O_3 is a Dr Jekyll, doing good, as we shall see shortly. But at ground level it is one of the evil greenhouse gases.* Along with CO_2, CFCs, halon and methane, it has the property of absorbing the heat which is radiated back from earth into the atmosphere. Since most ozone is produced from vehicle pollution, it is the one gas which can be eliminated effecti-vely. Vehicle exhaust can be cleaned up by using either catalytic converters in the exhaust system or lean-burn-engines. The converters eliminate the nitrogen oxides almost completely. Unfortu-nately, in getting rid of another of the exhaust gases, carbon monoxide, they produce CO_2. The monoxide is a poisonous gas for people, whereas CO_2 is not. In the sense of preventing harm to human health, the elimination of carbon monoxide is good. In the sense of adding to the greenhouse effect, the production of CO_2 is bad.

Lean-burn engines do emit small amounts of carbon monoxide and nitrogen oxides, but like the converters they produce CO_2. It is a case of "heads I win, tails you lose"! Vehicle exhausts, however "clean", have a bad effect. At the moment, some countries (but not the UK) con-trol the quality of vehicle emissions, usually by making compulsory the use of catalytic con-verters. Russia recently wished to export Samara cars to the USA. Before import licences could be issued this new range of vehicles had to undergo stringent emission-quality tests – which they passed, for the USSR has strict exhaust regulations itself.

Ozone at ground level is the equivalent of Mr Hyde. It is bad and we are creating it.

Ozone in the upper atmosphere is the equivalent of Dr Jekyll. It is good and we are destroying it.

Such is the present state of people's attitude to the environment and conservation. No wonder we are trying to persuade you, our readers, to consider conservation very carefully, so that later you can take sensible decisions based on knowing something about these complex prob-lems.

*At ground level ozone reacts with rubber to cause it to crack, it kills plants, such as the forest trees of Germany (see the book *Acid Rain* in this series), it causes lung irritation as one of the worst ingredients of urban smog, and it corrodes metal. Ozone has a strong pungent smell and a slightly blue colouring.

Invisible fumes rise from the vehicles on the motorway to add to the atmospheric pollution. Most visible exhaust is water vapour. The harmful gases cannot be seen.

Catalytic converters are made up of an outer metal case and an inner honeycomb structure of expensive metals such as platinum.

November 1987. A policeman masked against the smog tells a Hamburg motorist that all cars have been banned from the streets until the weather changes.

The ozone layer

There seem to be few things more enjoyable for people who live in the cooler parts of the world than to lie in the sun, those with lighter skins seeking to become tanned to a bronze colour. Sun-seeking is the main reason why so many people put up with the delays and discomfort of cheap air travel on their way to Spain and other Mediterranean holiday resorts. California, in the USA, and Australia conjure up pictures of long sunny days on golden beaches. But – there always is a "but" – too much sun shining on fair skins can cause illness, from the simple nausea people suffer on holiday with burning of the skin, to very severe damage as with skin cancer, called melanoma. (Melanin is the pigment in the skin which colours it to various shades of brown.)

It is the ultra-violet radiation (*see page 60*) within the sunshine which does the damage. Fortunately for us, most of the ultra-violet rays emitted by the sun do not reach us here on earth. 99 per cent of these rays which bombard our planet are soaked up by a narrow region of OZONE encircling the world about 15 miles above our heads. Even here the ozone forms only one part in every 100,000 parts of the atmosphere.

As explained on page 42, oxygen (O_2) is made up of two oxygen atoms linked together. If another atom is added, ozone (O_3) is created. Ozone is always quite ready to release this third atom and to return to being O_2; ozone is said to be chemically unstable. Remove the O_3 from the atmosphere and you remove the earth's protection from ultra-violet radiation, with possibly devastating results for human life.

In the early 1970s there was great concern about the effect of supersonic aeroplanes, such as Concorde and military aircraft, on the ozone-rich areas of the lower stratosphere in which they flew. The nitrogen oxides and water vapour from the planes' exhausts destroy the ozone at this high altitude where the air contains little oxygen. This concern no longer hits the headlines, probably because the development of such aircraft has been limited and because a far greater danger has been shown to exist.

The names of Sherry Rowland and Mario Molina ought to be widely known. As so often happens, some scientists achieve fame, and

If all of the ozone in the atmosphere were collected together, it would form a layer no thicker than a £1 coin.

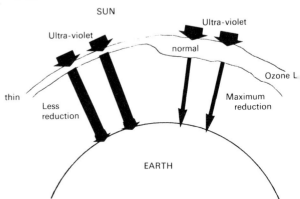

The effect of a reduction in the thickness of the ozone layer. The thicker the ozone layer, the greater the filtering out of the ultra-violet rays.

others, perhaps more deserving, remain in obscurity. In 1973 these two chemists published the results of their investigations into the effects of chlorofluorocarbon gases (CFCs) on ozone and the ozone layer. They discovered that CFCs are broken down by ultra-violet light and that when this happens they release free chlorine atoms. These free atoms react with O_3 and are six times more effective than nitrogen oxides at destroying ozone. The scientists discovered that each chlorine atom would start a "chain" reaction, eventually breaking up thousands of ozone molecules. CFCs are the main enemy of ozone.

Since the manufacture of CFCs was big business, there was much opposition from the chemical industry to the findings of Rowland and Molina. Further research suggested that the problem was even worse, in that the destruction of ozone by CFCs is two or three times more than the scientists stated originally. The real problem is that CFCs released at ground level are not broken up by rainfall or any of the chemicals in the atmosphere. Over the years after their release they just float up into the sky to await their breakdown by the sun's ultra-violet rays.

The most recent research, particularly over

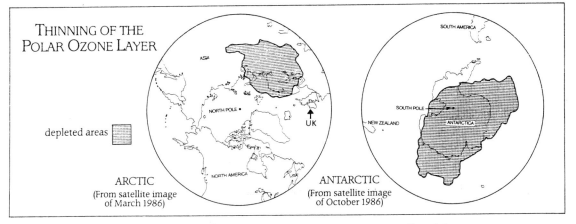

THINNING OF THE POLAR OZONE LAYER

depleted areas

ARCTIC
(From satellite image
of March 1986)

ANTARCTIC
(From satellite image
of October 1986)

In recent years, large thin areas have been observed in the ozone layer over both of the earth's poles.

(Source: Environment Canada).

the South Pole, has shown that here, at least, the ozone layer has already been affected. Since 1957 the amount of ozone has fallen by 40 per cent. During the Antarctic spring (autumn in the northern hemisphere) an area with little ozone in it has begun to appear. This is the "hole in the ozone layer" much publicized in the press and on TV in recent times.

Because ozone affects the heat balance of the earth, climates could change. What can be done to limit the damage to the ozone layer before it is too late?

A Dobson spectrophotometer in use at the British Antarctic Survey base. It is used to measure the amount of ozone in the air above the station.

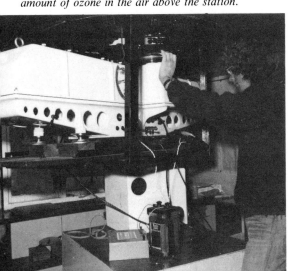

How CFCs destroy the ozone.

An O3 molecule is formed by a free O atom joining an O2 molecule	free oxygen atom + oxygen molecule → ozone molecule
A CFC molecule with 3 Cl atoms loses one Cl atom because of UV radiation	UV radiation, chlorofluorocarbon molecule → free chlorine atom
The free Cl atom takes one O atom from an O3 molecule to form ClO and O2	free chlorine atom + ozone molecule → chlorine monoxide + oxygen molecule
When the ClO meets a free O atom it loses it to form O2 and leaves the Cl atom free to destroy another O3 molecule	chlorine monoxide + free oxygen atom → free chlorine atom + oxygen molecule

---

An alternative explanation for the depletion of the ozone layer

Dr David Harper, a scientist of the Agricultural and Food Science Centre in Belfast, claims that a quarter of the chlorine in the upper atmosphere could come from wood-rotting fungi. The chlorine reacts with ozone to break it chemically and thus thin the protective ozone layer. It is estimated that $5\frac{1}{2}$ million tonnes of natural chloromethane is produced annually, much of it from fungi. These are widely distributed in both the temperate and the tropical parts of the world.

Down with CFCs!

Something *must* be done to prevent CFC gases going up. They are something which we can fairly have a "down" on. In fact, it is not a case of keeping them down but of eliminating their use as far as possible and, where they are still necessary, of controlling the industrial processes so that they cannot escape into the atmosphere.

It is obvious that an aerosol spray actually ejects CFC into the air, along with the liquid which it is intended to propel. The use of CFCs in aerosols has been banned in Canada and the USA for several years. Canada, in particular, has taken the lead in showing concern about the ozone layer. That country acted as host to an international gathering at Montreal in September 1987. From this came what is known as the Montreal Protocol (see page 57).

Unfortunately, European countries, especially the UK, have not been as responsive to the depletion of the ozone layer. On January 18th, 1988, the matter was debated in the British Parliament. During the course of the debate it was suggested that:

Aerosol cans are beginning to show if they are free of CFC gases. Keep a look out, too, for a new type of aerosol which contains no propellant gases. It consists of a thin plastic bottle inside a rubber cover. When the bottle is filled with a product at a high pressure the rubber and the plastic bottle expand. The stretched rubber tries to contract to its original size so that, when the escape valve at the top is pressed, the squeeze effect of the rubber forces the product out into the air. No gas is used, so there is no damage to the ozone layer.

You may have noticed that TV and other advertising for some solid products, particularly body deodorants, have been taking a positive line about preventing damage to the ozone layer. This is a good development because the advertisement is not only selling the product but making the viewer aware of the CFC problem.

1. The Montreal Protocol, due to be effective on January 1st, 1989, should contain tougher measures to combat CFC use.
2. More research should be carried out by the UK, in addition to the investigation work by the British Antarctic Survey, the Meteorological Office, the Atomic Energy Research Establishment and the Science and Engineering Research Council.
3. Cans containing CFC should be labelled.
4. Using non-essential CFC should be stopped.
5. Restrictions ought to be placed on vehicle air-conditioning systems.
6. Non-ozone gases should be used in fire extinguishers.

Only a month later, eight toilet goods makers agreed to stop using CFCs. They were Colgate-Palmolive, Beechams, Carter Wallace, Cussons, Elida Gibbs, Gillette, L'Oreal and Reckitt/Colman. This still left nearly a third of aerosol makers to join them. Even so, in their booklet *The Aerosol Connection*, the Friends of the Earth include a 26-page-long list of CFC-free aerosol sprays.

In Britain the CFC gases are made by ICI. In October 1988 the company announced that a complete phasing-out of the use of CFCs needed to be agreed internationally, and that a £50 million research programme to develop "ozone-friendly" substitutes was under way at its Runcorn laboratories. Some test products were being made.

Prince Charles has given a lead to us all, in our everyday lives. He has banned the use of CFCs in his own home, as it is possible to find other methods of carrying out the usual domestic tasks without them. We will have to rely on commerce and industry to control the use of CFCs in other processes. IBM UK Ltd use a CFC gas called Propaklone as a solvent to clean their computer boards of too much flux (used for soldering components). By careful control they now recycle 99 per cent of the chemical,

preventing it from escaping into the air. They are investigating an alternative, so that they can stop using Propaklone completely. However, substitutes can have their own problems. For instance, butane, used as a propellant instead of CFC, has the disadvantage of being highly inflammable.

Old refrigerators pose a problem. Their cooling systems contain CFC. In times past old 'fridges' have been broken up for their scrap metal with the CFC gas escaping into the air. One supplier of refrigerators BEJAM now collects the old ones and takes care to remove the CFC safely. This will need to be done by all companies and new refrigerators will have to contain non-CFC coolants. This is an enormous task.

In some cases it is practical and possible to change back to an older system. After using CFC-made plastic egg boxes for some time, the Co-operative Wholesale Society (the Co-op) has switched back to using cardboard containers made from recycled paper. 4,500 shops no longer using CFC boxes must surely have an effect on manufacturers and encourage them to cut down on the use of CFCs in their processes.

Perhaps the most important thing is that we are all, at last (CFCs were invented in 1928), considering the conservation implications of these gases. Not only will their decline help prevent further ozone damage; it will also reduce the Greenhouse Effect.

FRIENDS of the EARTH

THE AEROSOL CONNECTION

your guide to saving the ozone layer

5th Edition

The EEC and CFCs

In March 1989 the European Economic Community, in the words of Lord Caithness, one of the UK's environment ministers, made a "quantum leap forward" by agreeing to ban the use of CFC gases almost completely by the year 2000. In fact the UK government is striving to eliminate CFCs well before that. President Bush also announced a similar policy for the USA.

Until all cans are labelled to show the propellant gas being used, it is necessary to refer to the list of "safe" aerosols in the Friends of the Earth's publication.

Methane

Have you noticed a slightly sickly sweet smell coming from your dustbin or pile of rubbish? It is not a smell coming from any one particular thing, just a general dustbin-related aroma. Perhaps you have read reports in the press of explosive gas building up underground, where rubbish has been tipped and covered with soil. The smell and the gas are both caused by the production of methane, a highly inflammable gas. Methane is so easy to set alight that it can be used as a cooking gas or to run electric generators. At many former rubbish tips the gas is collected and piped to a nearby factory for making power.

Methane is also given off by rice paddy fields. A paddy is a piece of ground surrounded by slightly raised banks which form a watertight compartment. The paddy is flooded and rice plants are pushed down into the mud at the bottom. In the hot and wet conditions, vegetable matter decays in the water. "Swamp gas", which is methane, is given off.

Flatulence is the polite name given to gas created in the stomachs of animals during the digestive process. We are no different from other creatures where this is concerned and, after certain foods especially, our tummies may well

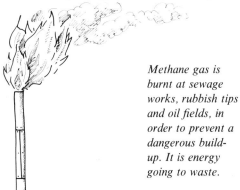

Methane gas is burnt at sewage works, rubbish tips and oil fields, in order to prevent a dangerous build-up. It is energy going to waste.

start to make gurgling noises as the gas inside is produced. The four-stomach digestive system of cattle is particularly "suited" to the production of gas, and rumblings from their insides are common enough noises in a cattle shed. Each cow makes a lot of gas which passes out into the air. The gas is, as I expect you have guessed, methane. If the manure of the cattle (or of any

Methane gas build-up 'is no danger'

Methane gas is building up at seven rubbish dump sites in Shropshire, but the county council said today there is no danger to residents.

"As far as I'm aware there is no danger. We are continually carrying out work so if there is any build up of methane gas it can escape quickly into the atmosphere," said county waste disposal officer Mr Gary Williams.

SHROPSHIRE STAR
AUGUST 11, 1988

The danger of explosion from methane is recognized by the Shropshire waste disposal officer – but to control this by polluting the atmosphere is to ignore the Greenhouse Effect. An important energy source is being frittered away.

other animal) is collected it is called slurry. By pouring the slurry into a closed container more methane will be made. This can be collected and used as a gas fuel, as with the rubbish tips. Nowadays one gas is known commercially as biogas. Firms have been established to exploit this and to sell biogas tanks for installation on farms.

I would not be writing about methane if it were not one of those gases which absorbs heat. It is a greenhouse gas and there is much evidence to suggest that it is increasing in quantity. It is a most unfortunate fact that large areas of rainforest are being cleared to make space for cattle ranching. Burning the vegetation produces CO_2. Removing the vegetation means there is less photosynthesis to absorb the world's overproduction of CO_2. Ranching cattle on the cleared ground, ultimately for their hamburger meat, leads to the production of more methane gas. So there is a threefold increase in the Greenhouse Effect.

Every effort should be made, therefore, to collect whatever methane is being produced, so as to prevent its transfer into the atmosphere. This can be done with the rotting rubbish in the tips. It can be done with the cattle slurry to produce biogas as a fuel. But it is not possible to deal with the methane produced in the rice paddy fields, and it is highly unlikely in these times of grain shortages in the developing world that such a staple crop as rice will be reduced.

Heavy bulldozer compacting the rubbish

ipes alread in position

At Packington Hall, Coventry, rubbish is being tipped into old sand and gravel workings and also on the ground surface to make an artificial hill. At all stages of the tipping, methane-collecting pipes are placed into the rubbish so that the gas may be drawn off easily and efficiently. Ultimately, the gas will power a turbine to produce 3.7 MW of electricity.

The wet conditions of growing rice encourage the production of methane gas.

A farm waste treatment plant at a monastery in Northern Ireland. Cattle waste is converted into a source of heat and a rich organic fertilizer.

Cattle shed

Control kiosk

Gas storage bag

Digester

El Niño

Christmas 1982 and New Year 1983 were not a happy time in many places in the Tropic regions of the world. In Australia bush fires raged and the small town of Macedon was destroyed by flames. Dust-storms of topsoil darkened the skies, and thousands of sheep had to be slaughtered in the worst drought of this century. In Indonesia, the Philippines, India, Sri Lanka, Southern Africa, Central America and Mexico, lack of rain ruined the crops. In the Bantu areas of South Africa, 45,000 domestic animals were slaughtered. In many places people starved. Yet in California and the Western States and Gulf States of the USA, in Cuba, Ecuador and Peru, torrential rainstorms brought flooding with huge landslides. In Salt Lake City in Utah, mud slides blocked many of the streets. The Mississippi River was only just prevented from flooding and gorging a new course to the Gulf of Mexico. Hurricanes hit the Hawaiian islands, and in Tahiti 25,000 people were made homeless. In the Philippines, half a million people lost their homes after Hurricane Vera struck. Five hundred tornadoes blasted Texas in a few months.

Drought and flood and wind; starvation and homelessness; fire and dust-storms; all these, it seems, were caused by a reversal of the equatorial current across the Pacific Ocean, creating a tongue of warm water 8000 miles long. A rise in water temperature of 14°F (7.7°C) above the normal affected the weather system, with the destructive results already listed. EL NIÑO had struck!

This is the name given to the abnormal warm water which comes at Christmas time, the time of the coming of the Christ Child. The Spanish *El Niño* means "The Child". This change in ocean temperature occurs at irregular intervals. There have been eight in the last 40 years, but that of 1982/83 proved to be so intense that it caused a major dislocation of the world's largest weather system – that of the equatorial regions.

El Niño is responsible for extreme changes of weather. Scientists are able to explain the way it works and to relate changes in local climatic conditions to it. But what triggers this phenomenon about every four or five years, and what made the 1982/83 El Ninõ so devastating? This they cannot explain. Perhaps, after all, it is just part of a bigger change in global climatic conditions.

One of the catastrophes: on February 16th, 1983, bush fires raged in SE Australia. At least 71 people died. Here, burnt-out cars and destroyed buildings mark the devastation in the town of Macedon.

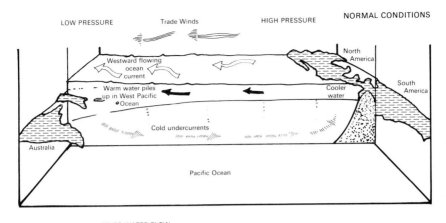

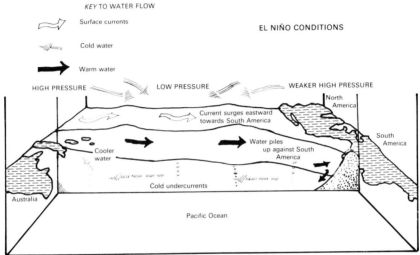

Cross-sections of the Pacific Ocean show how El Niño alters the direction of ocean currents.

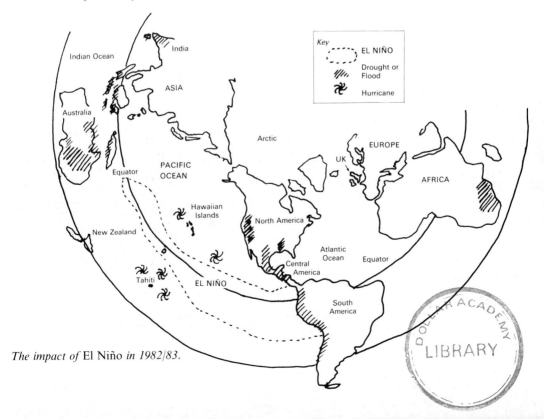

The impact of El Niño in 1982/83.

Sunspots

Climate change – another explanation?

At the time they called it the Little Ice Age. It lasted for many years until the end of the seventeenth century, 300 years ago. In London the River Thames froze over during the winters and a blanket of snow persisted through many months before and after Christmas. Pieter Breugel and other painters became well-known for the snow scenes which they painted at that time, including pictures of people playing and skating on the Thames and the Dutch canals.

Scientists believe that the average temperatures over Europe were then 1°C lower than usual. This does not sound very much, but as I have said in connection with the warming of *The Greenhouse Effect*, only a small average change can have serious consequences. As to the explanation for the lower-than-normal temperatures of the Little Ice Age, it has been observed from the records of astronomers that there were no sunspots on the sun at that time. Indeed, no sunspots were seen for over 50 years, which is a very unusual occurrence. The light from the sun is so intense that it seems to us to be the same brightness all over. However, astronomical observations show the sun's surface to vary in colour and the variations are attributed to variations in temperature. Darker patches on the sun are referred to as sunspots, where it is estimated that the temperature is about 2000°C cooler than elsewhere. These darker areas last for only a few weeks, but then others appear to replace them.

There are several theories for the appearance of sunspots. The one most favoured nowadays is that they are caused by strong local magnetic fields reaching the surface of the sun.

Sunspots seem to appear in cycles: over a number of years they grow in number to a maximum and then decline to a minimum before returning to another maximum. On average, the cycle takes 11 years, although it has varied from 7 to 17 years. Double cycles of 22 or 23 years have also been established, with every other 11-year cycle being seen to be more intense. Some scientists claim to have established an 80-year cycle and even an extreme 400-year cycle. Such conclusions are possible because observations of

Part of an oil painting by Abraham Hondius, The Frozen Thames, *1677. With London Bridge in the background, people can be seen enjoying themselves on the frozen Thames as well as using the ice to cross the river. Every year, during the "little ice age", ice fairs were held, with fortune tellers, side shows, pig roasts and the like.*

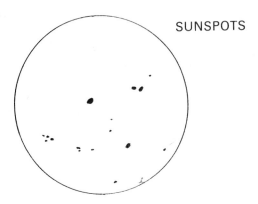

SUNSPOTS

In full sunlight, viewed through a reflecting telescope, dark spots can be seen on the surface of the sun. The pattern changes as time passes.

the stars, of which the sun is one, have been long recorded in world history.

Some climatologists associate sunspot activity with the variations in world weather. At present, we are in an 11-year cycle which began with a maximum of sunspots in 1980, saw a minimum in 1985/86 and is heading towards another maximum in 1991. It is claimed that the complete surface of the sun has become hotter and brighter than normal, so that the darker spots now are as bright as the brighter parts were before. The amount of sun reaching some places in the world will therefore vary and, it follows, so too will the weather conditions. These meteorologists explain the drought in North America by saying that in the late 1980s we are experiencing a period of doubly intensive sunspot activity. Unfortunately for the farmers of the grain-growing areas of the American mid-west, according to this theory, conditions are going to become drier until 1992, after which it will be less dry for five or six years. The weather experts who support the "sunspot theory" claim that similar conditions produced the Dust Bowl in the mid-1930s, and also maintain that other climatic changes worldwide are attributable to these sun surface variations.

Sunspots do affect the weather directly. Since we derive all of our energy and weather conditions from the presence of the sun, it makes sense to believe that variations in the intensity of the light and heat on its surface do affect our climate. That a real, but small, variation in weather is associated with sunspots is generally agreed. What is *not* accepted by most meteorologists is that sunspot activity can alter world climate by such a significant amount as to explain the changes attributed to the Greenhouse Effect.

Volcanoes

Climate change – another explanation?

Few of us have had to experience a volcanic eruption close to. For one newspaper reporter, "a 6 mile high ... aerial blanket of ash and gas transformed the noon sky over the province into a murky half light as farmers around the volcano abandoned vast stretches of rice fields now lying under large layers of mud" (*Daily Telegraph* report of the Mayon volcano in the Philippine Islands, September 1984). We have already seen (page 29) that volcanoes contribute to the amount of CO_2 in the atmosphere and therefore to the Greenhouse Effect. It is also considered that they can affect our climate more directly. The "murky half light" of the Mayon volcanic eruption was a result of the sun being blocked out by the ash. But what of the volcanic material which is first shot into the sky when a new or reawakened volcano explodes? Much of the EJECTA (the name given to the lava and ash blown into the sky) falls quickly back to earth in the area near to the erupting volcano. Some reaches the higher atmosphere where the upper-air winds spread it to other parts of the globe. By its very nature, the ash will act as a barrier to the rays of the sun, preventing some of the heat reaching the land. The rays are reflected from the small particles back into space.

Falls of red dust onto sou h-west England are proof that dust rising from the ground is carried by the upper-air winds. The red dust comes from upwelling dust-storms in the Sahara Desert and lands in England on the shiny surfaces of cars and on the laundry on washing lines. The volcanic ash from Mount Krakatoa, after its eruption in 1883, was thought to have reached 34 miles high and to have remained in the upper air long enough to have encircled the globe several times.

Not all volcanoes erupt in the same way. The dust from St Helena (USA, 1980) was ejected at a low angle, so that it came quickly to earth, while that from El Chicón (Mexico, 1982) shot more directly into the upper air and so had the chance to spread around the world.

According to one scientific theory, recent changes in climate have not been a result of the Greenhouse Effect at all, but have been caused by the fact that there has NOT been any major volcanic eruption in the last two years. The last "big one" was an eruption in Colombia in April 1986.

Mount Vesuvius, near Naples in Italy, is constantly adding gases to the atmosphere, although full-scale eruptions are single events with many years in between.

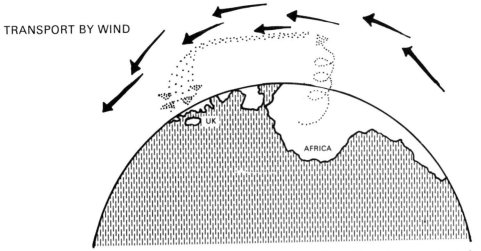

TRANSPORT BY WIND

UK

AFRICA

Violent storms over the Sahara Desert draw dust into the atmosphere. Blown by upper-air winds, it is deposited in Britain.

When volcanoes erupt, they may send their "ejecta" (gases, lava, volcanic bombs) out at a low angle to the ground or straight up into the atmosphere. With the low angle, there is more chance of the CO_2 and other gases being rained back to earth. High in the stratosphere they will remain longer, to add to the Greenhouse Effect.

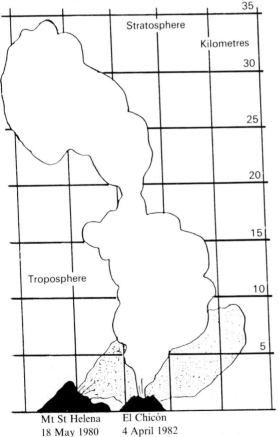

Stratosphere

Kilometres

35

30

25

20

15

Troposphere

10

5

Mt St Helena   El Chicón
18 May 1980   4 April 1982

It led to the deaths of 25,000 people. With no volcanic ash blocking the sun, more heat has reached the earth and a significant warming has taken place. Yet, to put all of the blame on the lack of volcanic activity seems to give too much importance to the volcanoes which do erupt. It is far more likely that the lack of volcanic eruption is just a further cause among many including the Greenhouse Effect, of our climate becoming warmer.

It is also significant that eruptions of the El Chicón type, which force volcanic material into the stratosphere, provide chemicals which react with the ozone layer. Even in years of no major eruption, an estimated average of 17 million tonnes of volcanic chemicals reach the upper atmosphere. Additions like the 500 million tonnes from El Chicón make for an increased danger to the ozone and a further thinning of the layer by natural causes.

We must not be too hasty to put all of the blame for climatic alterations at the door of CO_2 and CFCs.

Heavy grain harvest in 1986 and 1987

In 1986 and 1987, that is, in the two years following the Colombian eruption, the USA had bumper crops of corn. By 1988 the air was clear of this volcanic ash and the crops partially failed due to the drought (see pages 8-9).

Political aspects

It is only when you fly from coast to coast across the USA that you realize how big the country is. Each of the 50 states is different, with its own government and its way of doing things. Yet the people of all the states are drawn together by being part of one nation governed centrally from Washington and led by a President: politically they are one. In New York there is a large building outside of which fly the flags of most of the countries of the world, including that of the USA. This is the United Nations building. In no way are the countries which make up this organization politically united. Although much cooperation takes place there are many more matters on which members are not united.

The United Nations building, New York. *Can you see the flags of each of the United Nations flying outside their Headquarters?*

The United States of America has strict controls about pollution which apply everywhere in that country, although variations do occur between states. National control makes sense, because the smoke from a factory chimney in Philadelphia will be carried by the wind to St Louis, New Orleans or Denver, depending on which way it is blowing. Similarly it can spread to Toronto in Canada, or across the Atlantic Ocean to London. As these are different countries, no rules exist between them, unless special treaties are agreed by their governments. Air pollution does cross national borders; the fumes from the Drax power station in Yorkshire affect the atmosphere, which in turn affects the lives of people everywhere. The environment is not just a United Kingdom or a USA environment or the environment of any one country; it is a world environment – and people, including politicians, are beginning to realize this.

If you look at the back cover of this book you will see a quotation from the World Conservation Strategy. This is an attempt to get agreement between nations on conservation matters. Since the WCS was written, other reports about the future of the planet have been produced. *Our Common Future*, published in 1987, has become known as the Brundtland Report. It said that the "global commons", such as the atmosphere and the oceans, are parts of the natural environment which fall outside the control of individual nations and which are often undervalued, overused and abused. Wide-ranging and more effective laws – perhaps an international "Law of the Atmosphere" – are required to protect the air above us.

Political agreements are happening. In 1979 there was an international convention on Long-Range Transboundary Air Pollution and an "accord" was signed by 34 nations. In 1985 an agreement was reached to reduce sulphur emissions in each country by 30 per cent. For many reasons, three of the biggest producers of sulphur dioxide did not join the "30 per cent Club"; they were the UK, the USA and Poland. In 1985, too, a convention was held in Vienna for the Protection of the Ozone Layer.

sions had to be reduced beyond the aims contained in the protocol. Only Russia felt that more proof was needed that CFCs were the guilty gases. China, India and other developing countries were most concerned about the costs involved in new technology and demanded that the Western countries who manufactured the CFCs should be the ones to pay the price. This would be known as 'Phase Out Aid'. Other world climatic conferences are planned for the 1990s.

Behind many of these initiatives has been the United Nations Environment Programme (UNEP), part of the UN organization about which most people hear very little.

Mrs Gro Harlem Brundtland, *Prime Minister of Norway, and Director of the United Nations World Commission on Environment and Development, which produced the 1987 report* Our Common Future.

Several countries, including Canada and the USA, banned the use of CFC gases in aerosol spray cans. In September 1987 the Montreal Protocol (the first step towards a full agreement), asking for a 50 per cent reduction in the emissions of CFCs into the air by 1999, was drawn up, and by June 1988 31 countries had formally accepted the idea. In October 1988 the British government said that it would accept the protocol and improve its reduction in CFC production to 85 per cent by 1990. It also convened a meeting in London in March 1989 to bring together over 100 countries to discuss the problem of the destruction of the Ozone Layer.

At the "Changing Atmosphere World Conference" of June 1988, in Toronto, over 300 scientists, economists, environmentalists, legal experts and politicians emphasized the problems of the Greenhouse Effect and the destruction of the Ozone Layer. After the 1989 London Ozone conference the number of countries agreeing to the Montreal Protocol was increased to over 50. There was general agreement that CFC emis-

---

An important political occasion

Date   September 27th, 1988
Place   The annual dinner of the Royal Society in London
Speaker   Mrs Margaret Thatcher, Prime Minister of the United Kingdom

For the first time, a British Prime Minister spoke out about the danger that humankind may have "unwittingly begun a massive experiment with the system of the planet itself". She said that she doubted whether the implications of the hole in the ozone layer over Antarctica and its links with the gradual warming of the world climate – the "Greenhouse Effect" – had been fully appreciated. "We need to consider the wider implications for policy – for energy production, for fuel efficiency and for reforestation. ... We have an extensive research programme at our Meteorological Office and we provide one of the world's four centres for the study of climatic change. We must ensure that what we do is founded on good science to establish cause and effect".

Some comments after her speech

"Mrs Thatcher's speech ... places climatic change at the top of the worldwide political agenda." (Charles Clover, *Daily Telegraph*)

"The speech is an enormously interesting development. The importance of the Greenhouse Effect is at last getting through to world leaders." (Tom Burke, Green Alliance)

"Risks are associated with a do nothing and await results approach." (Jonathon Porritt, Friends of the Earth)

Atmospheric cures

Where our health is concerned, "prevention is better than cure". It is better to avoid becoming ill, if we can, than to face a cure afterwards. Therefore we put up with the unpleasantness of vaccinations and injections which will prevent us catching measles, mumps or tetanus. It is rather the same with the air above us. It would be far better to stop polluting it rather than having to try to clean up the mess afterwards. At the same time though, we could perhaps try to clean the atmosphere, using the same kind of technological skill which makes hospitals successful places for curing ailments in people.

Men and women have tried to alter the weather throughout history. Prayers are offered in churches when droughts occur. Rainmaking ceremonies are commonplace, such as those performed by the KOSHARES (medicine man) of the Pueblo Indians in New Mexico, USA, whose many tribal dances include one for rain. In other circumstances, the Koshares will be an ordinarily dressed US citizen.

MAN-MADE RAIN TAMES CHINA'S FOREST FIRE

THE worst fire in Communist China's history has been put out after threatening to spread across the border and join up with another fire in Russia.

Since Sunday nearly half an inch of rain has fallen on the Doxinganling Forest in north-east China, helping the tens of thousands of fire-fighters to win their 20-day battle.

The last two areas of the fire still alight had been advancing towards virgin forest areas in Inner Mongolia but they have now been extinguished by 'artificial rain.'

More than 3,000 shells of silver iodide were fired into the air and dry ice was spread in the sky to create rain.

The Daily Telegraph, May 27, 1987

Already we have said that using energy more efficiently will mean that there is less need for electric and other power. This in turn will reduce the burning of fossil fuel and the amount of extra CO_2 created. One technological cure, then, would be to pursue a world energy conservation programme aimed at using power resources in a more effective way and to investigate urgently the provision of electricity from alternative, non-polluting sources (wind, solar, tidal, wave, nuclear fusion, geo-thermal, biomass – all of which are described in another book in this series).

Some of the other ideas are almost science fiction. But many of the science fiction ideas of the past have become reality, so no idea must be rejected out of hand. The scientific skills of the year 2020 are likely to be as much ahead of now as the skills of today are ahead of those of 1950.

There are two main ways of applying a cure: (1) Consume the greenhouse gases; (2) Cool the temperature of the atmosphere by as much as the Greenhouse Effect is heating it up.

Consuming the gases

We have already discussed the way that plants absorb CO_2. Tree planting on a vast scale is one solution, in particular to replace the millions of hectares of rainforest which have been lost. A worldwide campaign to plant trees wherever possible would have an important effect and, at the same time, would increase people's awareness of the problem of world heating. Remember the coccolithophores and their appetite for CO_2? Might it be possible to cultivate enormous floating "forests" of them on the oceans? The problem of their sulphur by-product would need to be tackled. Other methods for the destruction of other greenhouse gases may be found.

Cool the temperature

It is estimated that a 2 per cent reduction in the amount of sunshine reaching the earth would compensate for the temperature rise. Every house could have white walls and white roofs, and this would increase the reflection of the rays of the sun back into space. More impractical is the idea of floating white polystyrene chips on

It is possible to cause rainfall artificially by dropping silver iodide and "dry" ice on clouds. This lowers the temperature, which in turn causes rain to fall. It is a very expensive method and does not work every time; the rain sometimes falls somewhere other than where intended (and paid for!). The cloud here has been "seeded" with chemicals. The hole is where the rain was formed.

the oceans to form a reflective cover. What an enormous task! But if it saved the earth from disaster it might be worth it. Sun-shields in space covering enough sky to block out 2 per cent of the sunshine, might provide another answer, though what about the places which would be in permanent shadow beneath? I suppose the shields could be moved around, but I imagine their location would give rise to bitter arguments between countries. Since we know that volcanic dust and chemical droplets reflect sunshine naturally, perhaps artificial particles could be sprayed into the air. One suggestion is that a fleet of 700 jet aircraft fly around the world every day, high in the stratosphere, spreading 35 million tonnes of sulphur dioxide every year. This would reflect the unwanted 2 per cent of the sun's power back into space. Might it not add to the acid rain and ozone layer destruction problems?

What of the depletion of the ozone layer? Might it be possible to project capsules of replacement O_3 into the sky? Space factories might orbit the earth, producing ozone to release into space. Or it might be possible to destroy the harmful chemicals in a sort of "star wars" way, with laser beams tuned to select the "bad" and ignore the "good".

It all sounds rather absurd? But so did the

idea of men walking on the moon earlier in time! One thing is certain. Worldwide agreement will be necessary.

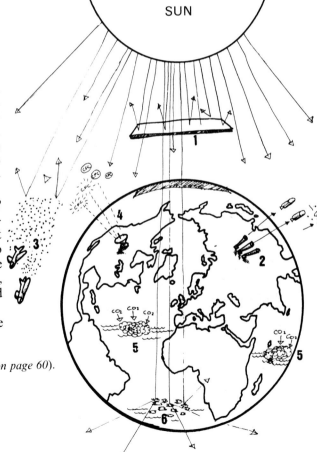

"Cures" for the problem? (*see the "Balloon story" on page 60*).
1 Satellite sun shield
2 Mortars shooting ozone capsules into space
3 Aircraft discharging gas particles
4 Laser beams exploding CFC gases
5 Algae blooms floating on sea
6 Polystyrene rafts reflecting sun's rays.

59

What can you do to reduce the Greenhouse Effect?

Take an interest in your home's electricity bill and help to reduce it by switching off unnecessary lights and heaters, not wasting hot water and using "off peak" electricity where possible.

Take a similar interest in the electricity consumption at your school. Ask your headteacher if you may form an electricity conservation group.

Talk to your parents, friends and neighbours about home insulation against waste of energy.

Check the aerosols which you have in your home against the approved list from the Friends of the Earth booklet. Stop using those which contain CFCs.

Check before you buy new aerosols that their propellant gases are "ozone-friendly". If in any doubt, buy non-aerosol products such as roll-on deodorants, hand pump sprays, polish in tins and bottled cleaners.

Stop having garden bonfires. Instead, make a compost heap of all your vegetation waste.

Keep a look-out for smoking chimneys. Try to persuade the owners to clean up their emissions.

Take photographs of bad offenders and send the pictures to your local Friends of the Earth group or even to the local newspaper.

Write to your local councillor/MP/Euro MP, to tell them how concerned you are about CFC gases and about the whole Greenhouse Effect problem.

But remember to be polite:

ask your parents' advice before taking any action
ask your teacher if it involves school

Whatever you do, be positive. Don't just sit back and grumble. Remember, it is *your* world and *your* environment as well as other people's.

Balloons may help in ozone rescue

Oswestry's Thunder and Colt, of Maesbury Road industrial estate, hopes to float hundreds of solar-powered ozone generators into the stratosphere in a bid to repair the ozone layer.

They will be lifted 17 kilometres above the poles where the protective shell is thinnest.

The generators are small versions of domestic ionisers, used to clear charged air, and will pump out ozone gas to replace some of the molecules which are being destroyed by chlorofluorocarbons.

Thunder and Colt's international sales manager, Mr Crispin Williams, said the balloons would be filled with helium to help them stay up for up to nine months.

Now the firm is busy designing a revolution system to keep the generators airborn by automatically dumping ballast as the helium gas slowly leaks away.

If the £50,000 pilot project is successful up to 100 balloons will be floated at each of the two poles to pump gas into the air.

Shropshire Star, 15 March 1989

The Effect of Ultra Violet Radiation

Ultra violet radiation can destroy DNA, the most important cell type in our bodies, sometimes known as the 'building blocks of life'. Ultra violet causes sunburn, eye damage, skin cancer, snow blindness and skin wrinkling. It affects plants preventing the germination of seeds, including those of agricultural crops and trees, slowing down photosynthesis and killing off algae. Since algae are important in the seas it may well be that fishing, too, will be adversely affected by ultra violet radiation.

Glossary

Acid rain Rain which is a weak acid with a pH value below 5.0. The term is used for acid mist, acid hail, acid snow and even for "dry" acidic deposition.

Aerosol Small drops of liquid which float in the air.

Air conditioning A system for cooling the air in closed areas.

Air pressure The weight of air pushing down on the earth. In weather forecasts it is referred to as "low" or "high". It is measured in millibars.

Algae Very small plants which often float on water to give it a green colour.

Altimeter An instrument for measuring height.

Atmosphere The area which surrounds the earth, made up of gases and water vapour.

Biogenic effect The way plants affect the environment.

Charcoal Wood which has been "cooked" and which becomes a substitute for coal.

Climate The average weather conditions for any place.

Climate model A computer prediction of future climate.

Cloud Droplets of water floating in the atmosphere.

Cloud seeding Dropping silver iodide crystals onto cloud to make it rain.

Condensation Water changing from a gas (vapour) to a liquid when it becomes colder.

Development programme A plan to alter an area, hopefully for the better. Usually a richer country aiding a poorer one.

Dust Bowl An area of wind erosion where the top soil is blown away. The mid-west area of the USA became known as the Dust Bowl in the 1930s.

Evaporation Water changing from liquid to gas (vapour) when it becomes hotter.

Fossil fuel Something which has been formed millions of years ago, brought out of the ground and burnt. Once used it has gone for ever. Coal, oil, natural gas, peat and brown coal are all fossil fuels.

Insulate Prevent heat loss by putting in a barrier such as glass fibre "wool".

Montreal Protocol An international agreement signed in 1987 to limit the use of CFC gases.

Nomadic Describes a way of life where people have no fixed home but move from place to place.

Nuclear fission Splitting the atom to create heat.

Nuclear fusion Joining atoms together to create heat.

Nuclear power Using the heat from a nuclear reaction to create electricity.

Nutrient That part of an animal's or plant's intake which feeds its growth.

Off-peak electricity Electricity which is available more cheaply when demand is low, as at night.

Photosynthesis The process by which a green leaf makes starch and sugars for food. It results from the action of sunlight on the plant. CO_2 is absorbed during the process.

Phyto-plankton Minute algae (green plants) floating on top of the sea. Zooplankton (small creatures) "graze" on it.

Radioactivity The process of production of harmful radiation when atoms are split.

Salination The deposition of salt on the land after flooding.

Slash and burn A method of clearing forest areas by cutting and burning, usually by peasant farmers.

Sorghum A grain crop used for animal feed.

Transpiration The loss of water from plants.

UN The United Nations. Its headquarters are in New York.

Resources list

Reports

The Heat Trap – threats posed by rising levels of Greenhouse Gases, Climate Research Unit of the University of East Anglia, published by Friends of the Earth, 1988

Into the Void? A report on CFCs and the Ozone layer, Kathy Johnston, Friends of the Earth, 1987

Books

Atmosphere, Oliver Allen, "Planet Earth" series, Time-Life Books, 1983

The Cousteau Almanac, Jacques-Yves Cousteau, Columbus Books, 1981

Gaia Atlas of Planet Management, Editor Norman Myers, Pan Books, 1985

The hole in the sky, John Gribbin, Corgi, 1988

Magazine articles

"El Niño", *National Geographic*, Vol. 165 No. 2, February 1984

"Carbon Dioxide and World Climate", *Scientific American*, August 1982

Colour transparency packs

Planning for survival, *Acid rain* and other packs, International Centre for Conservation Education

Books in the "Considering Conservation" series

Acid Rain, Philip Neal, 2nd edition, 1988

Disappearing Rainforest, Robert Prosser, 1987

The Encroaching Desert, Norman Farmer, forthcoming 1990

Energy, Power Sources and Electricity, Philip Neal, 1989

Farming and Food Supply, Derrick Golland, 1988

War on Waste, Joy Palmer, 1988

The World's Water, Joy Palmer, 1987

Useful addresses

Central Electricity Generating Board,
Department of Information & Public Affairs,
Sudbury House,
15 Newgate Street,
London EC1A 7AU

Department of the Environment,
43 Marsham Street,
London SW1P 3PY

Environment Canada,
4905 Dufferin Street,
Downsview,
Ontario
M3H 5T4
Canada

Friends of the Earth,
26–28 Underwood Street,
London N1 7JQ

Greenpeace,
36 Graham Street,
London
N1 8LI

International Centre for Conservation
    Education,
Greenfield House,
Guiting Power,
Cheltenham,
Gloucestershire

National Coal Board,
Hobart House,
Grosvenor Place,
London
SW1X 7AE

United Nations Environment Programme,
PO Box 30552,
Nairobi,
Kenya

United States Environmental Protection
    Agency,
215 Fremont Street,
San Francisco,
California 94105
USA

Index